AF774542

Fractal Modelling: Growth and Form in Biology

Jaap A. Kaandorp

Fractal Modelling

Growth and Form in Biology

With 149 Figures, 27 in Colour
Foreword by P. Prusinkiewicz

Springer-Verlag Berlin Heidelberg GmbH

Jaap A. Kaandorp
Department of Computer Science
University of Amsterdam
Kruislaan 403
1098 SJ Amsterdam
The Netherlands

ISBN 978-3-642-63402-4

Library of Congress Cataloging-in-Publication Data
Kaandorp, Jaap A., 1958– Fractal modelling: growth and form in biology/
Jaap A. Kaandorp: with foreword by Przemyslaw Prusinkiewicz p. cm.
Includes bibliographical references (p.) and index.
ISBN 978-3-642-63402-4 ISBN 978-3-642-57922-6 (eBook)
DOI 10.1007/978-3-642-57922-6
1. Growth-Computer simulation. 2. Morphology-Computer simulation. 3. Fractals
4. Computer graphics. I. Title. QH511.K28 1994 574.3'0113–dc20 94-44611 CIP

Originally published by Springer-Verlag New York Berlin Heidelberg New York in 1994
Softcover reprint of the hardcover 1st edition 1994

Cover Design: Design & Concept, Heidelberg
Typesetting: Data conversion by Springer-Verlag

45/3140 – 5 4 3 2 1 – Printed on acid-free paper

For Sarita and Mikael

Foreword

The relationship between growth and form is one of the most exciting problems in biology. The complexity of developmental processes that transform a seed into an adult tree or a fertilized egg into an animal is difficult to comprehend and defies traditional mathematical descriptions. Their limitations led Benoit Mandelbrot to the discovery of fractals: the intricate geometric objects more suitable for representing irregular forms of nature than figures of Euclidean geometry. Mandelbrot observed that many fractals could be obtained using a strikingly simple construction invented in 1905 by Helge von Koch and consisting of repetitive substitutions of given geometric figures by sets of other figures. In 1968, Aristid Lindenmayer proposed a similar mechanism as a mathematical model of the development of multicellular organisms. In this case, cell divisions were viewed as substitutions of the mother cells by their children. The analogy between the substitution of geometric figures and the division of cells related fractals to developmental biology.

In this book, Jaap Kaandorp applies mathematical models and computer simulations rooted in fractals to explore the relationship between growth and form in marine sessile organisms: corals and sponges. The sophistication of the models progresses from simple geometric abstractions to comprehensive models of specific organisms found in nature. Commendably, Kaandorp emphasizes the predictive power of the models as the essential criterion of their practical value. One interesting application is biomonitoring, in which a mathematical model is used to establish the relationship between the shape of an organism and its environment. This relationship makes it possible to use the shape of a growing organism as an indicator of environmental conditions. The purpose may range from pollution control to the study of long-term climatic changes.

Most of the book is devoted to the description of Kaandorp's original results obtained in the scope of his Ph.D. research at the University of Amsterdam, followed by a fellowship at the University of Calgary.

Formally educated in both computer science and biology, Kaandorp displays a profound knowledge of the living organisms he describes and the computer science techniques he needs to devise and implement the models. The book collects the results accumulated to date and presents a vibrant account of science in progress.

Calgary, January 1994 *Przemyslaw Prusinkiewicz*

Acknowledgements

The work described in this book could be completed thanks to the support and contributions of many persons. I would like to mention some of them especially.

The main part of the work in this book was done at the Department of Computer Science of the University of Amsterdam. The biological part of this project was done in cooperation with biologists from the Institute of Taxonomic Zoology of the University of Amsterdam.

Many valuable comments on the manuscript were made by Frans Groen and Edo Dooijes from the Department of Computer Science, their support in this research has been very important. I am thankful to Hans Lauwerier from the Department of Mathematics. The discussions with him on fractal geometry and his computer demonstrations were very inspiring.

I wish to thank several persons from the Institute of Taxonomic Zoology of the University of Amsterdam. Jan stock made several important remarks on the manuscript. Rob van Soest gave many valuable comments on the biological relevance of the models. The discussions with him about sponges and corals were very useful for me. The previous work of Wallie de Weerdt on the Haplosclerida (sponges) and *Millepora* (hydrocorals) was a source of inspiration. She kindly provided one of her drawings to be used in this book (see Fig. 5.8) and three of her underwater photographs from the Caribbean (Figs. 3.4, 3.6, and 3.7). The black and white photographs are all made by Louis van der Laan. His skill in photography and advice were very important in this research. I owe very much to Mario de Kluijver. He provided an experimental basis for the models and made the two under water photographs shown in Figs. 3.2 and 3.16. Thanks to him the field experiments described in Chap. 4 could be carried out. I hope we can continue this fruitful cooperation in future.

An important part of the work on the 3D models was done at the Department of Computer Science of the University of Calgary in Canada under a Government of Canada Award. In Calgary I had the opportunity to work in an excellent computer graphics environment, among people with much experience in modelling biological objects. I would like to thank Przemyslaw Prusinkiewicz for his support and all the interesting discussions we had. I wish to thank Mark Hammel, Deborah Fowler, Larry Aupperle, Kees van Overveld, Camille Sinanan and Brian Wyvill for all their helpful comments and the nice time at the University of Calgary.

I would like to thank Paul ten Hagen from the Centre for Mathematics and Computer Science in Amsterdam for offering me the opportunity to work on computer graphics and fractals. The period that I worked at this centre was an important part of my education before starting the research for my book.

I owe very much to my parents Truus and Jaap Kaandorp who always supported me.

Many others contributed to this book: Annette de Gee, Rob Bakker, Marcel Wijkstra and Behr de Ruiter provided parts of the visualization software. Lisanne Aerts collected the special growth form shown in Fig. 6.2, Erik Meesters provided the section of *Montastrea annularis* shown in Fig. 3.9, Freerk Hiemstra showed me many microscopical sections of sponges, Zbigniew Struzik knew an answer to many of my questions during the writing of this book, Jean Pierre Boon kindly permitted me to use the picture in Fig. 2.1, several of the pictures in Chap. 5 were visualized with the program *rayshade* written by Graig Kolb, much of the research was done on computer equipment made available by IBM under ACIS contract. I would like to thank Hans Wössner from Springer-Verlag for the pleasant cooperation.

Finally, I would like to thank especially Sarita, with whom I always could discuss my work; without her support I could never have written this book.

Amsterdam, January 1994 *Jaap A. Kaandorp*

Table of Contents

1 Introduction

In living organisms an almost infinite multitude of forms is found, yet there is still very little understanding how these forms emerge. The emergence of forms in the growth process of biological objects is one of the most fundamental problems in biology. The view that growth and form are interrelated has a long tradition in biology. A classical study on this subject is D'Arcy Thompson's (1942) book *On growth and form*. In this study the form of an organism is considered as an event in space-time and not merely a configuration in space. This view is also the basis for many of the mathematical models which have been developed to obtain insight into the morphogenesis of biological objects.

A mathematical model which has been used frequently to model biological pattern formation is known as the reaction diffusion mechanism (Turing 1952). This model describes diffusing chemicals, which can produce steady-state heterogeneous spatial patterns under certain conditions. In this theory of morphogenesis, patterns or structures result from this spatial pattern (the prepattern) of non-homogeneous chemical concentration distributions.

A well-known mathematical model for biological pattern formation is the L-system (Lindenmayer 1968). This model has recently been applied on a wide scale in computer graphics for the synthesis of biological objects. Some examples of L-systems will be discussed in Chap. 2.

Many objects in nature, in contrast with man-made objects, show at first sight a high degree of irregularity, non-smoothness and fragmentation. These objects cannot easily be described using traditional modelling techniques using spheres, lines, circles, etc. They do not resemble the "normal" objects of euclidean geometry. Closer observation reveals that these objects are often characterized by the remarkable property of self-similarity within a certain interval of scales: an enlargement of the object will often yield the same details. A well-known example of self-similarity in biology is the human lung. The bronchi and bronchioles form a tree-like branching pattern, where the branching of the airways on a smaller scale looks like the branching pattern at larger scales (Goldberger et al. 1990).

Constructions of sets with the property of self-similarity have been known in mathematics for a long time. These sets were often used as examples for which certain mathematical properties cannot be determined. They were formerly considered as pathological cases. In the book *The fractal geometry of nature* by Mandelbrot (1983) it is demonstrated that these self-similar sets are very useful objects, applicable as mathematical models for many objects in nature. Also in this book, the name "fractals" is coined for these self-similar sets; in the next chapter a more precise definition of fractals will be given.

Fractals

The study of growth and form in nature has been stimulated considerably by the development of fractal theory (Falconer 1990). The fractal quality can be demonstrated for many biological objects, for example blood vessel systems (Turcotte et al. 1985; Wlczek et al. 1989; Family et al. 1989), coral reefs (Bradbury and Reichelt 1983), and vegetation (Morse et al. 1985). Fractals often seem to serve quite well as mathematical models of biological objects.

DLA model

An important development in fractal theory was the modelling of growth patterns with the "Laplacian" (Niemeyer et al. 1984) models, which started in physics with the diffusion-limited aggregation (DLA) model (Witten and Sander 1981). Laplacian models are very successful in physics and can be generalized to describe many fractal growth phenomena (Sander 1986). Examples of the DLA models will be shown in Chap. 2.

In spite of the many studies, still very few exact results are available about the problem how forms emerge in the biological growth process. In almost all cases, it is not yet known how the genetic information is physically translated into the actual form. Much research is being done in biology, experimental as well as theoretical, in order to reveal more about the physics and mathematics behind the growth process in which the DNA code gives rise to certain shapes and patterns in the physical environment.

Simulation of growth and form

New developments in mathematics, physics, and computer science offer possibilities for biologists to obtain a deeper understanding of the emergence of form. With recent developments in computer science, it has become possible to carry out simulation experiments in which the growth process, the interaction between cells or skeleton elements, can be imitated in virtual computer objects. The capabilities of computer simulations are still too limited to simulate the complete growth process on a molecular level. It is even hard to imagine that it will ever be possible to carry out a complete simulation experiment in which a DNA molecule generates new molecules, where the generated molecules form cells, and in which cells finally interact in clusters. A comparison between the possibilities of the DNA in the cell and computer simulations is given by Murray (1990): "An idea of the immense complexity of a cell is given by comparing the weight per bit of information of the cell's DNA molecule, around 10^{-22}, to that of, say, imaging by an electron beam of around 10^{-10} or of a magnetic tape of about 10^{-5}. The most sophisticated and compact computer chip is simply not in the same class as a cell."

The desired level of refinement of the model

The first crucial step in the development of simulation models is the choice of the level of the elements which are interacting in the physical environment in a growth process. This choice will, firstly be determined by the desired refinement of the model and, secondly for practical reasons by the physical limitations of the computer hardware. An obvious choice in a simulation model of a seed plant, serving to yield an understanding of how the growth form develops, is the cellular level. A lower level, for example the atomic level, compared to the level in which the genetic information is encoded, would yield no better insight. A practical choice in the simulation of the form of a (stony) coral, a bryozoan, a sponge or a virus could be the level of the corallite, a zooid, a skeleton element (a spiculum) or a molecule, respectively. These elements are the typical basic building elements for these organisms which determine the final growth forms, for an exigent part.

Basic building elements

Modules

In a simulation model of the growth process of a vegetation of seed plants, a quite practical choice could be to simplify the level of the basic building elements to that of the modules (the apical meristems, see Harper et al. 1986). A still higher level is used by Koop (1989), where in a simulation system of forests the vertical projections of crowns and profiles of trees are used as basic elements in the model. In order to create a simulation model of a flock of starlings, it is probably not useful to descend down to the molecular level. A typical characteristic of growth processes, vegetations or communities of organisms is that they exhibit a behaviour which cannot be deduced from the individual composing elements (see Simons 1969). By some authors this is considered as a characteristic of life: "All collections of living things show properties unexpected from a knowledge of a single one of them" (Lovelock 1988). From the DNA of the starling it is probably not possible to deduce the final shapes of the flock of starlings. Together the basic elements exhibit a new, often highly complex behaviour. To obtain deeper understanding of these complex systems, simulation models will be often the only way available.

The choice of the research taxa is another crucial step in the development of simulation models. For example, simulating the growth process of seed plants on a cellular level may already lead to highly complex models with a vast number of parameters. Especially when the abiotic terrestrial environment is taken into account, a model simulating the growth of interacting cells will become too complex in number of parameters. The predictions which can be made with such a model are often within the range of normal fluctuations which occur in the real objects. This is a notorious problem in models of ecosystems: these often exhibit a highly complex dynamics and are often characterized by a limited predictability and low applicability in new situations, in spite of the high levels of precision used in the computation. In Saris and Aldenberg (1986) these models are even indicated as "artefacts of precision". The same problem can be encountered in for example economical, meteorological, and climatological simulation models.

Predictive value of the models

In order to develop a morphological model based on low-level elements, an attractive choice can be sessile marine organisms. The abiotic marine environment is characterized by a remarkable uniformity when compared to the terrestrial environment. Many of the environmental parameters influencing the growth process of sessile marine organisms, such as salinity, oxygen, and sometimes even temperature may be assumed to be constant. For many marine organisms the physical environment can often be reduced to two key parameters: water movement and light. For these reasons, in the marine environment important simplifications can be made in modelling the physical world, compared to the terrestrial or the freshwater environment.

Marine sessile organisms

A second important simplification which can be made in modelling for a large group of sessile marine organisms is based on the fact that they exhibit a relatively simple growth process. The growth process of, for example, a (stony) coral, in which only the surface of the colony is alive and where new layers are deposited onto the dead core, is much simpler to model than the growth process of a seed plant. In spite of the relatively simple growth process, a remarkably high diversity of forms can be observed in a coral reef.

For these reasons, many of the examples shown in this book involve marine organisms. The intention of this book is not to discuss modelling techniques for marine organisms only. The modelling techniques developed should be considered as a basis from which to develop more complex models suitable for simulating other organisms, organs of organisms, and communities. In the last chapter some examples of these more complex models will be shown.

Coral reefs

One argument which still should be mentioned, in particular for investigating sessile marine organisms, as found in a coral reef, is that it has an important environmental application. Coral reefs form an important part of the ecosystems on earth. However, together with the rain forests the coral reefs belong to the many endangered ecosystems on earth. Yet very little insight has been obtained in these ecosystems. Since 1986 bleaching of corals has been observed in reefs along the coasts of Australia, Caribbean, and the Indo-Pacific Ocean, a phenomenon which may lead to dying of corals. This phenomenon is connected by some authors to the global warming of the earth (see Bunkley-Williams and Williams 1990; Glynn and Croz 1990; Goreau and Macfarlane 1990; Jokiel and Coles 1990).

Experimental work

In this book an attempt is made to demonstrate that there should be a clear relation between the real and the virtual objects. The development of a simulation model should be supported by experimental work. A simulation model which has not been tested against reality is in danger of lacking practical use. It is necessary to correlate the model with observations of the actual objects and experiments in order to verify all assumptions made in the model. Building simulation models is a typically interdisciplinary endeavour in which mathematics, computer science, biology, and experimental work are interwoven. During the construction of growth models it is necessary to describe the various aspects of the growth process in formal terms. This formal way of describing the process leads to a systematic approach, where none of the aspects can be neglected. The formal description together with the resulting simulation model can indicate which field experiments could deliver interesting results. Even if the model appears to be incorrect, this working method may lead to interesting new results.

1.1 Structure of the Book

In Chap. 2 several approaches to model forms are discussed and the advantages and disadvantages of the various approaches are compared. Some of the mathematical models mentioned above will be discussed in more detail. The development of a robust 2D geometric modelling system is described, which is then applied in the following chapters. It is demonstrated how from simple rules, stepwise more complex rules can be built. The geometric modelling system will be suitable for simulating a growth process in 2D. The modelling of a growth process of a simple (artificial, non-biological) object is discussed.

In Chap. 3 the same modelling method is used to model biological forms in 2D. As a case-study a certain growth process found in marine sessile organisms, such as sponges and stony corals, is used.

In Chap. 4 the development of methods to compare virtual and real objects is described. In this chapter experiments with real objects are discussed to verify the model.

In Chap. 5 the geometric modelling system of Chap. 2 is extended to 3D. A method is presented for modelling in 3D the growth process discussed in Chap. 3.

The subject of Chap. 6 is the application of the 2D and 3D models discussed in the previous chapters. Examples are given of how simulation models can be used in ecological research.

2 Methods for Modelling Biological Objects

In this chapter several methods for modelling biological objects are discussed. The methods described in this chapter have the potentiality to serve as morphological models of biological objects. In the first section a model for pattern formation, based on diffusing chemicals, is described. In Sect. 2.2 the iteration processes and fractals which form the general base of the methods described in the Sects. 2.3, 2.4, 2.5 and, 2.6 are discussed. In the last section of this chapter a review is given of the methods mentioned in the chapter and arguments are given as to which method is the most applicable for morphological models of growth processes.

2.1 Reaction Diffusion Mechanisms

One of the oldest mathematical models used for modelling biological pattern formation is known as the reaction diffusion mechanism (Turing 1952). This model describes diffusing chemicals, where often a system of two antagonistic chemicals is used consisting of an activator and an inhibitor. This model can be described as a system of equations (Murray 1990) in the form:

$$\begin{aligned}\frac{\delta A}{\delta t} &= F(A, I) + \mathcal{D}_{\mathrm{A}} \nabla^2 A \qquad (2.1)\\ \frac{\delta I}{\delta t} &= G(A, I) + \mathcal{D}_{\mathrm{I}} \nabla^2 I\end{aligned}$$

In these equations A and I represent respectively the concentrations of the activator and the inhibitor. The functions F and G represent the reaction kinetics and the right terms in both equations the diffusion process, where $\mathcal{D}_A$ and $\mathcal{D}_I$ are the diffusion coefficients for the activator and inhibitor. The diffusion process can result in a heterogeneous spatial prepattern of chemical concentration distributions, in the case $\mathcal{D}_I$ is much larger than $\mathcal{D}_A$. In Turing's proposal for a theory of morphogenesis, patterns or structures result from this prepattern. This prepattern can be defined as a heterogeneous spatial pattern of inhibitor and activator concentrations, while the resulting pattern can be considered as the realization of the prepattern. This model is especially suitable for generating patterns that may result more or less directly from this prepattern. It has been applied for simulating patterns found on shells (see Meinhardt and Klingler 1987 and 1992; Fowler et al. 1992), and coat patterns (Murray 1988 and 1990). In a mammalian coat pattern the hair colour is determined by the pigment cells, the melanocytes, which can produce pigment (melanin). It is believed that whether or not the melanocytes produce melanin is determined by the presence or absence of chemical activators and inhibitors (Murray 1988). Although these chemicals are not known yet, it is supposed that coat patterns are a reflection of an underlying chemical prepattern. It is possible to simulate such a prepattern, which is caused by diffusing activators and inhibitors; an example of such a simulation is shown in Fig. 2.1 (after Boon and Noullez 1987). This pattern was simulated in a two-dimensional lattice, where the cells can be in two states: activated (black) or inhibited (white). (Details about simulating a diffusion process in a 2D lattice will be discussed in Sect. 2.4) Many general and specific features found in mammalian coat patterns and shell patterns can be explained with this theory, although the theory itself still has to be confirmed by experimental observations.

Shells and coat patterns

Fig. 2.1. Simulation of a prepattern caused by diffusing activators and inhibitors in a 2D lattice. The cells in the lattice can be in two states: activated (black) or inhibited (white) (after Boon and Noullez 1987).

2.2 Iteration Processes and Fractals

The common basis of the methods discussed in the rest of this chapter is an iteration process (see Fig. 2.2). These methods have the potentiality to serve as morphological models of biological objects. In this iteration process the output of one iteration is used for the next one. The relation f between input and output may be linear or non-linear. The examples shown in the following sections are based on a linear relation. Examples of objects generated in a process with a non-linear relation are the Julia sets (see Julia 1918; Mandelbrot 1980 and 1983; Peitgen and Richter 1986). In these examples the relation f in the iteration process is a quadratic mapping in the complex plane. The objects which are generated in this iteration process are often fractals (examples of fractal objects will be shown later in this chapter). The process may also deliver normal geometric figures or single points of attraction, or may not converge. This depends on the choice of f in the iteration process.

An example of a linear relation in the iteration process is the construction shown in Fig. 2.3. In this construction the iteration process starts with a square (the initiator). In each iteration step an edge of the object is re-

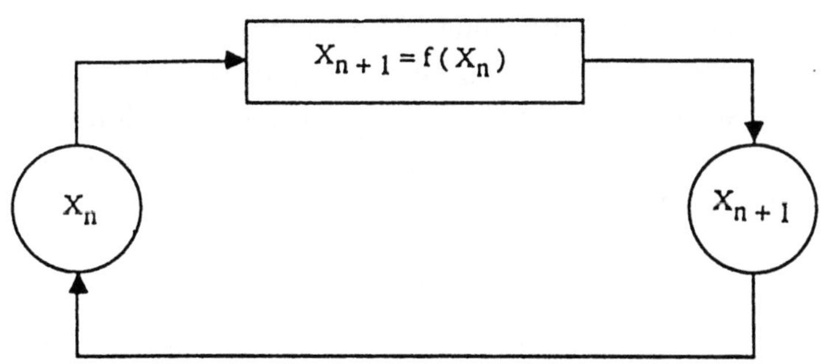

Fig. 2.2. Diagram of an iteration process in which the output of one iteration is used as input for the next one

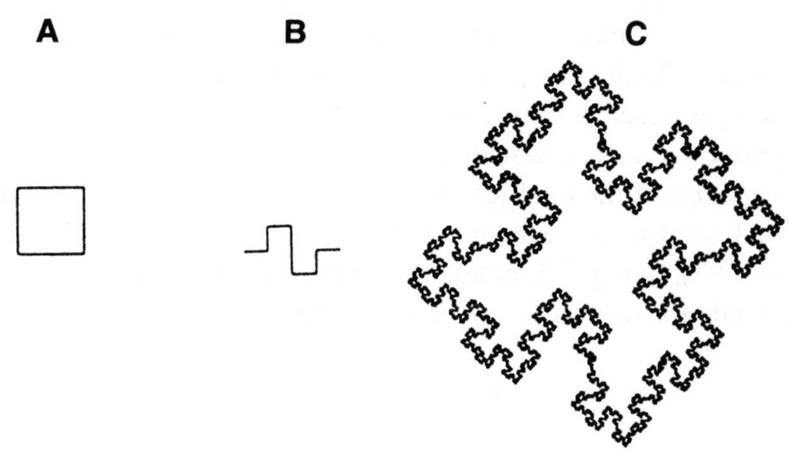

Fig. 2.3. Geometric construction of the quadric Koch curve: the construction starts with the initiator (A) and each edge of the object is replaced by the generator (B) in each iteration step. The iteration process results in the quadric Koch curve (C).

placed by a set of 8 edges (the generator). In each step a combination of a geometric scaling, rotation and, translation is done for each edge; together this combination can be written as a linear transformation in the iteration process. The process results in the curve shown in Fig. 2.3C, which is known, in the limit case, as the quadric Koch curve (Mandelbrot 1983). This type of curve was regarded in the past as a pathological case for which certain mathematical properties cannot be determined.

Koch curves

The Koch curve is characterized by three remarkable properties: it is an example of a continuous curve for which there is no tangent defined at any of its points, it is locally self-similar on each scale, an enlargement of the object will yield the same details, and the total length of the curve is infinite. The quadric Koch curve is an example of a fractal object. In Mandelbrot's book, fractals are defined as sets for which the Hausdorff-Besicovitch dimension exceeds the topological dimension. Fractals may be defined as sets with, in most cases, a fractional dimension, which is often indicated as the fractal dimension. Fractals show a self-similar structure, and this phenomenon may be used as the guiding principle. For a more general definition of fractal dimension see Hutchinson (1981), Dekking (1982), Hata (1985); Falconer (1985 and 1990) and Barnsley (1988).

Fractal dimension

The value of the fractal dimension D can, for this special case, be determined analytically: the value is 1.5 exactly. The value of D may be calculated for this self-similar curve made up of N equal sides of length r using (2.2) from Mandelbrot (1983). The ratio r of the length of a side of a fractal approximant and the preceding fractal approximant and is also known as the similarity ratio.

$$D = \log(N)/\log(1/r) \tag{2.2}$$

The value of the fractal dimension can also be determined experimentally. When an estimation of the total length is made by covering the curve with an equal-sided polyline with side length ϵ, the total length of the curve $L(\epsilon)$ increases when ϵ decreases, however without converging to a finite limit. The process in which the total length is estimated with an equal-sided polyline, where the length of ϵ decreases in successive approximations, is visualized in Fig. 2.4. The relation between the approximation of the total length $L(\epsilon)$ and ϵ is given in (2.3) (Mandelbrot 1983, see for approximation methods also Rigaut 1991, Dooijes and Struzik 1993). In Fig. 2.5 some estimations of the total length of the Koch curve, made for various values of ϵ, are depicted.

$$L(\epsilon) \sim \epsilon^{1-D} \tag{2.3}$$

The exponent D in this equation, in the case of the Koch curve, is the value of the fractal dimension (Mandelbrot 1983). The value of D can be estimated from Fig. 2.5, which yields a value $D \approx 1.5$. D can be determined analytically only in a few special cases. In general (for example for a biological object) D can only be determined experimentally, as shown in Fig. 2.5. Many biological objects are non-deterministic fractal objects, which are (statistically) self-similar within a certain interval of scales. Within this interval, fractals can be used as a mathematical model of the biological object.

Statistical self-similarity

An iteration process is a very natural way to describe growth processes in biology or in physics. In a growth process the last growth stage will

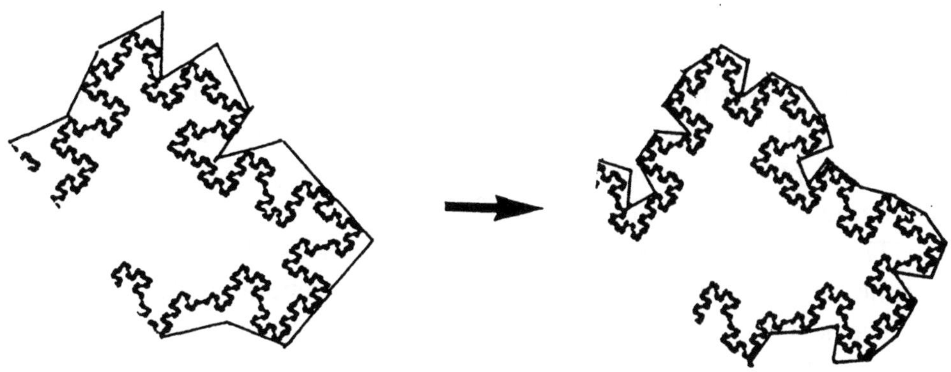

Fig. 2.4. Estimation of the total length of the curve shown in Fig. 2.3C. The total length $L(\epsilon)$ is estimated by covering the curve with an equal-sided polygon with side length ϵ; in successive approximations the length of ϵ decreases.

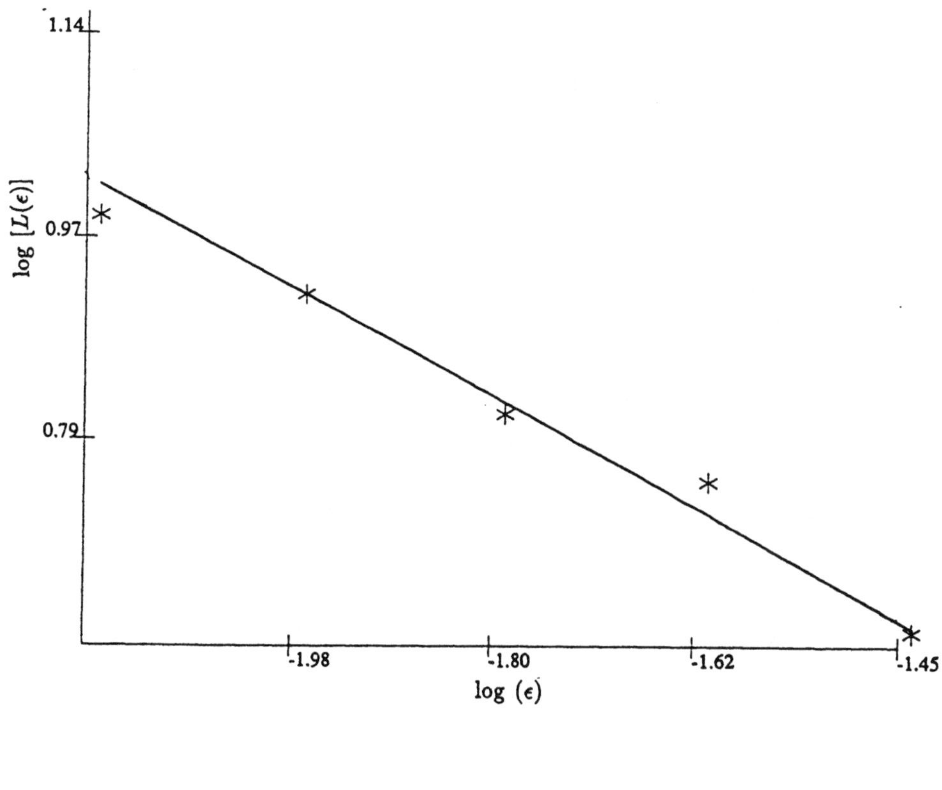

Fig. 2.5. Relation between the total length $L(\epsilon)$ and ϵ, where the value of $L(\epsilon)$ was estimated with the method displayed in Fig. 2.4

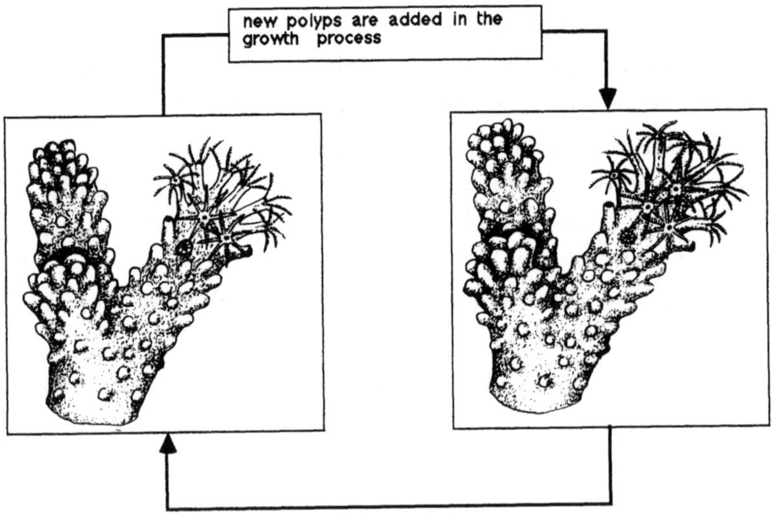

Fig. 2.6. Diagram of the growth process of *Alcyonium glomeratum*: the growth form of the preceding stage is used as input for the formation of the next growth stage.

serve as input for the next growth step. In Fig. 2.6 this iteration process is visualized for *Alcyonium glomeratum* (Octocorallia). In this organism new basic building elements, in this case the polyps, are added to the preceding growth stage in each growth step.

This iteration process is suitable for modelling the shape, as it emerges in time, in a growth process. The same process is suitable for modelling other aspects of growth, as for example the growth of a population. In this case, the size of a population is also determined by the size of the population in the preceding growth step.

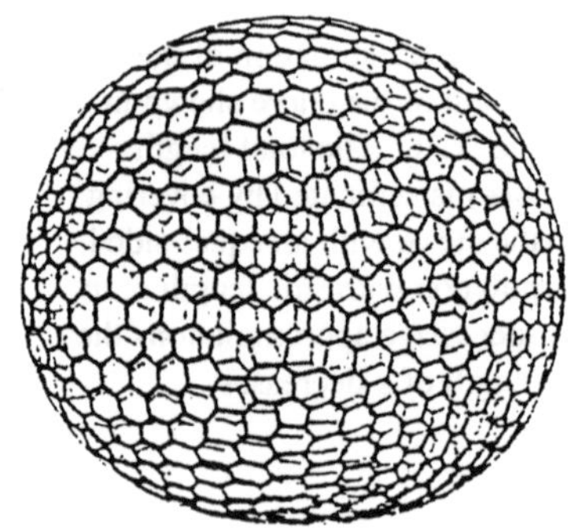

Fig. 2.7. Example of a spherical radiolarian, *Aulonia hexagona* (after Haeckel 1887)

As will be demonstrated, fractal objects result surprisingly often from iteration processes. The iteration process can be considered as the basis of growth processes in nature, which could be the explanation for the fact that fractal objects are so common in nature. It is indeed hard to find the objects of euclidean geometry in nature. Tetrahedron-shaped and cube-like organisms are hardly to be found. Sphere-like organisms can be found among *Orbulina* (Foraminifera) and radiolarians. An example of a spherical radiolarian *Aulonia* is shown in Fig. 2.7. Some more of the series of platonic solids (a nice and systematic description of these solids can be found in Wenninger 1971) such as the rhombic dodecahedron in Fig. 2.8 and the tetrakaihedron are quite common among cells. But fractal objects, like trees and clouds in the terrestrial world and branching corals and waves in the marine tropical world, dominate.

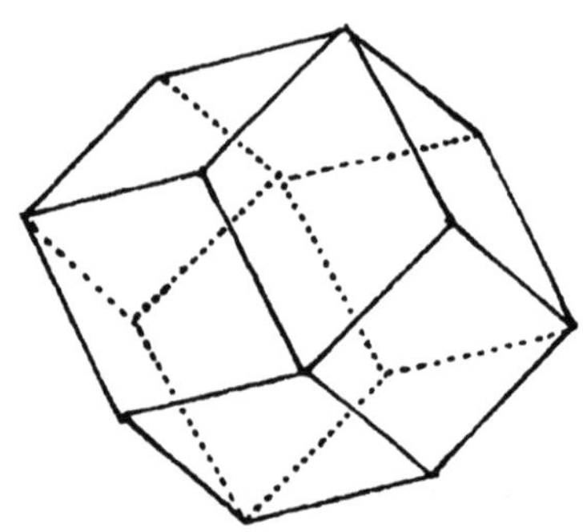

Fig. 2.8. Example of a rhombic dodecahedron, as can be found among cells

In many cases, as for example in many of the seed plants, the basic building elements, the cells, resemble in general the platonic solids of euclidean geometry, whereas they often aggregate into clusters with fractal characteristics. In some cases the basic building elements themselves are fractal objects; an example of this can be found among the Lithistids (Porifera), see Fig. 2.9, where the fractal spicula aggregate in cup-shaped, mushroom, or spherical forms.

Methods for modelling biological objects

The various methods for modelling biological objects, which are discussed in the following sections, differ mainly in the representation of the objects in the iteration process. In many cases these techniques can be considered as alternative approaches. They can also be combined with each other; an example of this will be given later on. The situation can be compared to the use of different data structures in computer science. It is often a fruitful approach to represent a problem in more than one type of data structure. One can then combine the benefits of the different representations.

2.3 Generation of Objects Using Formal Languages

Lindenmayer grammar

A well-known model for biological pattern formation, from botany, is the L-system or Lindenmayer grammar (see Lindenmayer 1968). The Lindenmayer grammar is similar to those known in conventional formal language theory as Chomsky hierarchy languages (see Hopcroft and Ullman 1979). A rather fundamental difference is that the rewriting rules (production rules) are applied simultaneously in the Lindenmayer grammar. In the L-systems strings are generated in an iteration process, as shown in Fig. 2.2. The symbols X_n and X_{n+1} in the iteration process are represented by strings in a formal language. The strings themselves do not contain geometric information; in order to translate the strings into

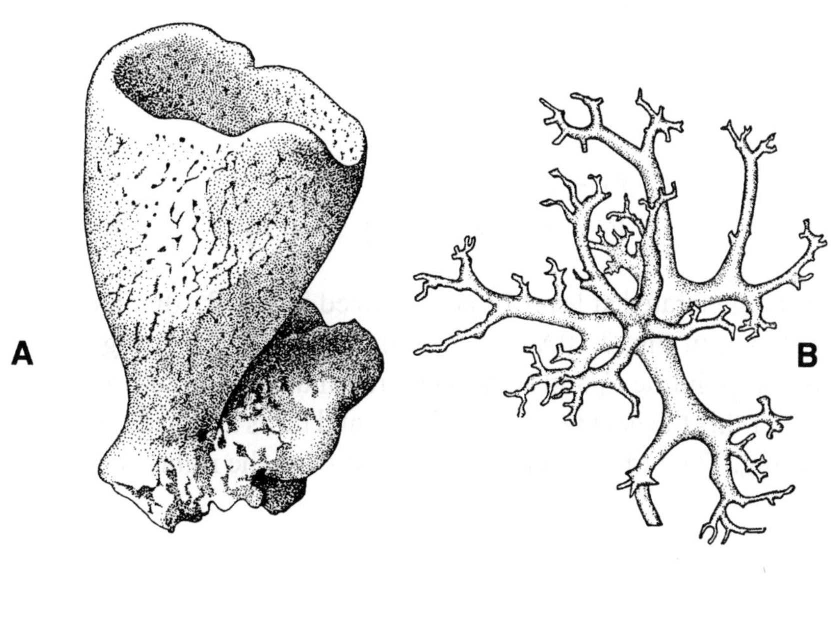

Fig. 2.9. A fractal-like basic building element (B) found in the Lithistids (Porifera, after Sollas 1878). The growth form which is constructed from these elements is shown in (A).

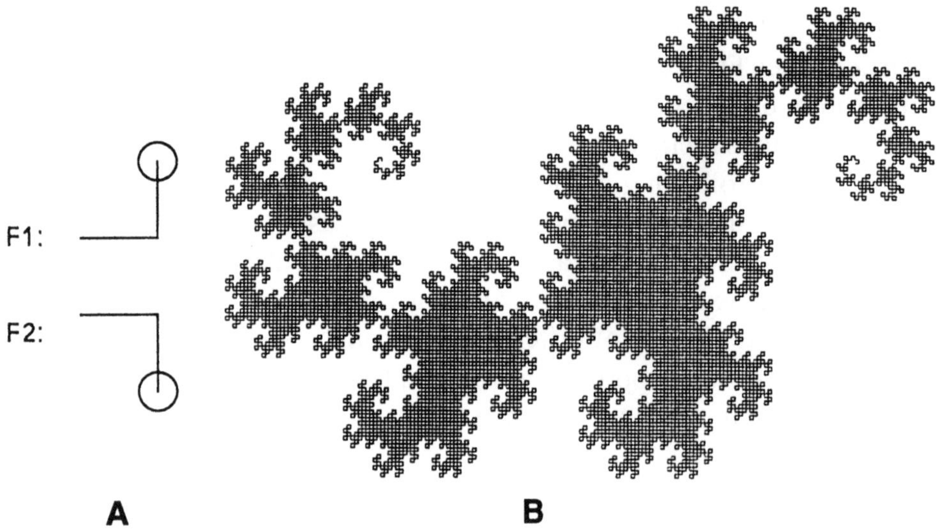

Fig. 2.10. (A) Drawing rule of the curve shown in (B) The symbols F_1 and F_2 are visualized as polygons, at the endpoint of each polygon (indicated as "O"), turns are made as indicated in the string. The symbol + is interpreted as a turn of 90° to the right and – as a turn 90° to the left. (B) Curve (Dragon sweep) resulting from the L-system depicted in (2.4) and the drawing rule in (A).

a morphological description, additional drawing rules are necessary. An L-system can be defined by using a triple K denoted by $< G,\ W,\ P >$, in which G is a set of symbols, W is the starting string or axiom, and P is the production rule.

Many of the "classical" fractal curves shown by Mandelbrot (1983) can be generated with L-systems (see Prusinkiewicz and Lindenmayer 1990). An example of such a fractal curve is the Dragon sweep, which can be denoted as: *Dragon sweep*

$$\begin{array}{rcl} K_{\text{dragon\_sweep}} & = & < G_{\text{dragon\_sweep}}, W_{\text{dragon\_sweep}}, P_{\text{dragon\_sweep}} > \\ G_{\text{dragon\_sweep}} & = & \{F_1, F_2, +, -\} \\ W_{\text{dragon\_sweep}} & = & F_1 \\ P_{\text{dragon\_sweep}} & = & \{F_1 \to F_1 + F_2, F_2 \to F_1 - F_2, + \to +, - \to -\} \\ < \text{iteration} > & & < \text{iterated string} > \\ 0: & & F_1 \\ 1: & & F_1 + F_2 \\ 2: & & F_1 + F_2 + F_1 - F_2 \\ 3: & & F_1 + F_2 + F_1 - F_2 + F_1 + F_2 - F_1 - F_2 \end{array}$$

The string generated at level 10 is visualized in Fig. 2.10B. The drawing rule is shown in Fig. 2.10A. In the visualization of the string one starts drawing the polyline F_1 and at the endpoint (indicated as "0") a 90° turn is made, after this the next polyline is drawn and turns are made as indicated in the string. The turns are indicated as + (90° to the right) and – (90° to the left).

L-systems and biological objects

L-systems have been applied for bio-morphological description (see for example Hogeweg and Hesper 1974). In De Boer (1989) examples of L-systems simulating division patterns in cell layers can be found. In this study the first five cleavage stages of a *Patella vulgata* embryo are simulated. In Frijters (1976) is shown by examples how L-systems can be applied to formalize the different florescence states of *Hieracium murorum*. In Renshaw (1985) is demonstrated how the root structure and canopy development of a sitka spruce *Picea sitchensis* can be simulated with this method.

An example in which L-systems are applied in bio-morphological description, is the generation of two different branching patterns: monopodial and dichotomous or sympodial branching. The two types of branching patterns are illustrated in Figs. 2.11 and 2.12. Both branching patterns may be defined as L-systems by using a triple $< G, W, P >$. In L-systems describing branching patterns, brackets are used to denote branches. The brackets represent a branch which is attached to the symbol left to the left bracket.

Monopodial branching

Monopodial branching may be represented by the following L-system:

$$\begin{aligned} K_\mathrm{m} &= < G_\mathrm{m}, W_\mathrm{m}, P_\mathrm{m} > \qquad (2.4)\\ G_\mathrm{m} &= \{0, 1, [,]\}\\ W_\mathrm{m} &= 0\\ P_\mathrm{m} &= \{0 \rightarrow 11[0]0, 1 \rightarrow 1, [\rightarrow [,] \rightarrow]\} \end{aligned}$$

< iteration >	< iterated string >
0 :	0
1 :	11[0]0
2 :	11[11[0]0]11[0]0
3 :	11[11[11[0]0]11[0]0]11[11[0]0]11[0]0

One possibility to visualize the generated strings for monopodial branching is to use the drawing rule shown in Fig. 2.11A. In this drawing rule the string 11[0]0 is visualized alternating as between the shapes a and b. The visualization can be described as a series of translations and rotations. The transformations (using homogeneous coordinates, see Foley et al. 1990) are represented by matrix operators, where $T(DX, DY)$ indicates a translation over the vector (DX, DY) and $R_P(\gamma)$ a rotation about an angle γ. When a coordinate frame is assumed, with the origin $O = (0, 0)$, each symbol in the string defines a transformation of the previous coordinate system A. The positions of the successive origins of the coordinate systems form the vertices used in the visualization of

Homogeneous transformations

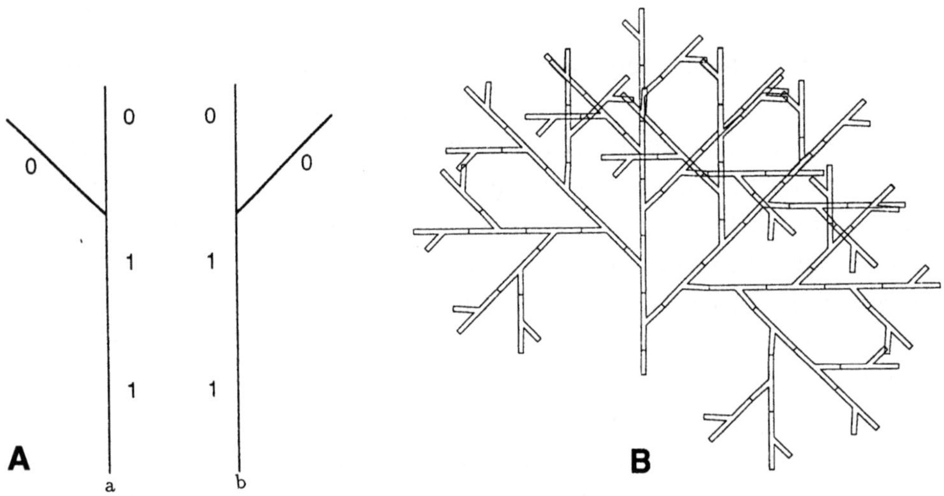

Fig. 2.11. (A) Drawing rule for monopodial branching: the string 11[0]0 is visualized as alternating between shapes a and b. (B) Visualization of a generated string, for monopodial branching, from level 6

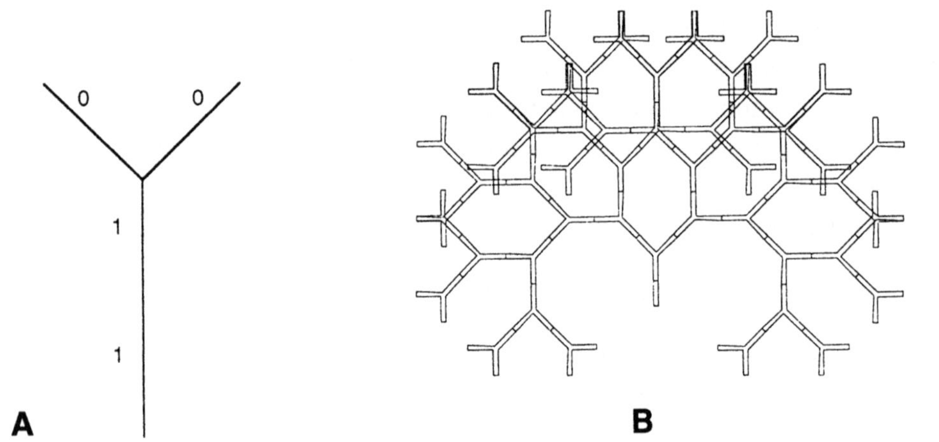

Fig. 2.12. (A) Drawing rule for dichotomous branching, visualization of the string 11[0][0]. (B) Visualization of a generated string, for dichotomous branching, from level 6

the strings. The visualization is done by drawing line segments between the successive vertices. The coordinate frame A, in a certain stage of the visualization, consists of the product of all previous homogeneous transformations. At the end of a branch the coordinate frame is reset to the one at the beginning of the branch. The visualization can be described in algorithmic form as:

$$\begin{array}{lll} 1: & A = & A \cdot T(\underline{a}); i = i + 1; \\ 0: & A = & A \cdot T(\underline{a}); i = i + 1; \\ [: & A = & A \cdot R(-\gamma) \text{ if } (odd(i)); \\ & & A \cdot R(\gamma) \text{ if } (even(i)); \\ & & stack(i); \\ ]: & & pop(i); A = A_i; \end{array} \tag{2.5}$$

where $\underline{a}$ is the vector $[0, DY]$ and $\gamma = 45^\circ$

The string generated at level 6 is visualized in Fig. 2.11B.

Dichotomous branching

Dichotomous branching may be represented by the L-system:

$$\begin{aligned} K_d &= \ < G_d, W_d, P_d > \\ G_d &= \{0, 1, [,]\} \\ W_d &= 0 \\ P_d &= \{0 \rightarrow 11[0][0], 1 \rightarrow 1, [\rightarrow [,] \rightarrow]\} \end{aligned}$$

$$\begin{array}{rl} < iteration > & < iterated\ string > \\ 0: & 0 \\ 1: & 11[0][0] \\ 2: & 11[11[0][0]][11[0][0]] \\ 3: & 11[11[11[0][0]][11[0][0]]][11[11[0][0]][11[0][0]]] \end{array}$$

The string 11[0][0] may be visualized using the drawing rule shown in Fig. 2.12A. The drawing rule uses the same translations and rotations as in Fig. 2.11A; in this rule both a rotation to the left and to the right are carried out. The result (from level 6) is visualized in Fig. 2.12B.

Many examples of applications of L-systems are from computer graphics, where they have been used on a wide scale for generating images of biological objects. Examples of these studies are: Aono and Kunii (1984); Smith (1984); De Reffye et al. (1988); Prusinkiewicz et al. (1988); Prusinkiewicz and Lindenmayer (1990).

In the examples shown in this section the final image can be generated by the interpretation of the generated strings, by applying the drawing rules. These strings can be expressed recursively. The strings generated in the system for monopodial branching (2.4) can be described recursively as:

$$S(n+1) = 11[S(n)]S(n) \tag{2.6}$$

L-systems and randomness

It is also possible to apply randomness in the production rules, for example in the following L-system, where a mixture of monopodial and dichotomous or sympodial branching is used:

$$\begin{aligned} K_r &= \ < G_r, W_r, P_r > \\ G_r &= \{0, 1, [,]\} \\ W_r &= 0 \\ P_r &= \{0 \stackrel{0.5}{\rightarrow} 11[0][0], 0 \stackrel{0.5}{\rightarrow} 11[0]0, 1 \rightarrow 1, [\rightarrow [,] \rightarrow]\} \end{aligned}$$

In this L-system the probabilities of applying the sympodial and monopodial production rules are indicated above the arrows. A string generated at level 7 is visualized in Fig. 2.13. In the case of this stochastic L-system the advantage of expressing the branching structures recursively is lost.

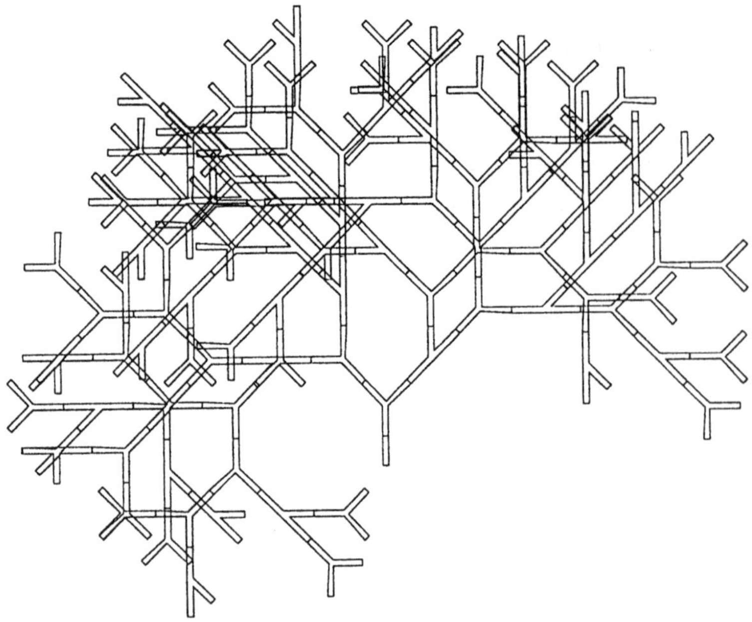

Fig. 2.13. Visualization of a generated string of level 7 in which randomness is applied

It is no longer possible to predict the string from a certain iteration level, as was done in (2.6) for a deterministic L-system. In the examples of L-systems shown above, the objects were generated in 2D, but it is also possible to extend these systems to 3D. Examples of this are given by Aono and Kunii (1984), Prusinkiewicz and Lindenmayer (1990).

L-systems and geometric restrictions

In L-systems it is not easy to introduce geometric restrictions in the iteration process. In the example of monopodial branching the geometric transformations from the visualization algorithm (2.5) are not included in the production rule of (2.4). A very obvious restriction in modelling growth processes is a rule which prevents intersections in the object. In Figs. 2.11B, 2.12B and 2.13 intersections occur everywhere in the objects. This limitation makes L-systems less applicable for developing a geometric model, where geometric restrictions are essential.

2.4 Diffusion Limited Aggregation Models

The Diffusion Limited Aggregation model of Witten and Sander (1981) has been used on a wide scale in physics for explaining various fractal growth phenomena, such as particle aggregation, dielectric breakdown, viscous fingering and electro-chemical deposition.

An example of fractal growth which can be described with this DLA model is a growing object (for example a bacterium colony in a petri dish) which is consuming a nutrient from its environment (see Meakin 1986). The concentration c is zero on the object and it is assumed that

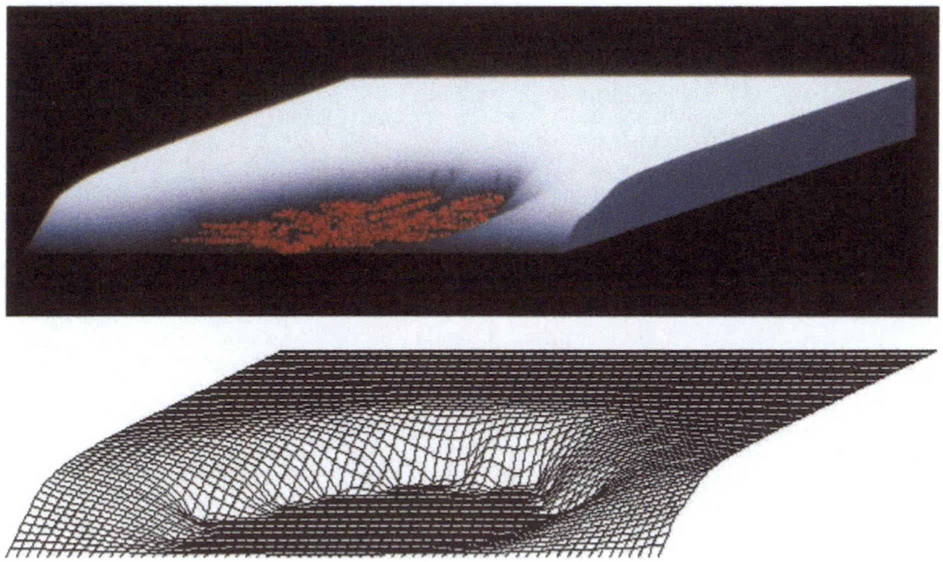

Fig. 2.14. Visualization of the meaning of the Laplacian model. In this figure a growing object (the object is displayed again in Fig. 2.15) is pressed into a sheet, which is fixed at the edges. The height of the sheet represents the local nutrient concentration; the object itself is situated on the bottom plane, where the concentration equals zero.

the diffusion process, described by (2.7) is fast compared to the growth process. In this diffusion equation $\mathcal{D}$ is the diffusion coefficient.

$$\frac{dc}{dt} = \mathcal{D} \nabla^2 c$$

The nutrient concentration is supposed to remain constant ($c = 1.0$) on a circle or sphere (the boundary), surrounding the growing object. The concentration field will attain a steady state, in which $\frac{dc}{dt}$ equals zero. The distribution of the nutrient concentrations around the object, in a steady state, is described by the homogeneous Laplace equation (2.7).

Laplace equation

$$\nabla^2 c \equiv \frac{\delta^2 c}{\delta x_1^2} + \frac{\delta^2 c}{\delta x_2^2} + \frac{\delta^2 c}{\delta x_3^2} = 0 \qquad (2.7)$$

$$\begin{aligned} c &= c(x_1, x_2, x_3) \\ c(x) &= 0 \text{ for } x \in object \\ c(x) &= 1 \text{ for } x \in boundary \end{aligned}$$

The meaning of the Laplacian model can be visualized by Fig. 2.14 (after Sander 1987). In this picture a growing object (the same object is shown again in Fig. 2.15) is pressed into a rubber sheet, which is fixed at the edges. The height of the rubber represents the local nutrient concentration. The concentration is maximal at the borders of the sheet, where nutrient is supplied continuously and is minimal at the object itself, where the nutrient is consumed. The steady state is described by the curved surface, which satisfies the Laplace equation. The object grows fastest at the sites

Fig. 2.15. DLA cluster generated within a 1000 x 1000 lattice, where 1.0 was taken for η in (2.8). The local concentrations are visualized as alternating black and coloured regions and the object itself is displayed in red. The concentration decreases in the coloured basins when the colour shifts from pink to blue.

where the highest nutrient gradients occur. These sites can be recognized in Fig. 2.14 by the steepest slopes in the rubber sheet. Intuitively it can be seen that growth is the fastest at the tips of the object, where the highest gradients occur.

Discharge patterns

In the previous example of the DLA model, c represents the nutrient concentration. With the same model, fractal growth patterns which arise in a dielectric breakdown can be explained. In this case, c in (2.7) represents the electric potential (see Niemeyer et al. 1984) and the growing object is represented by a discharge pattern (known in the literature as a Lichtenberg figure) on which the electric potential is zero on the pattern and 1.0 on a circular electrode. The same DLA model is used to simulate flow velocity in a Hele-Shaw cell (see Feder 1988). These cells are used in physics for experiments in hydrodynamics. In this case the c in (2.7) represents the flow velocity and this model can be applied to model fluid instability phenomena, known in the literature as viscous fingering (see Nittmann et al. 1985; Feder et al. 1989)

Growth in non-equilibrium

The Laplacian model can be generalized to describe other fractal growth phenomena (see Sander 1986; Stanley and Ostrowsky 1987), for example growth of electro deposits (see Brady and Ball 1984) and particle aggregation, where growth takes place in non-equilibrium. For a growth process where a cluster of particles is formed and in which the cluster grows by adding new particles, growth in equilibrium can be defined as a process where in the cluster formation the most stable configuration is formed and the particles are allowed to change sites in order to achieve this stability. In a growth process in non-equilibrium the possibility that particles change sites is limited, the consequence is that a cluster is formed which, in most cases, is not the most stable configuration.

Growth in equilibrium

In a growth process in equilibrium, for example as found in many growing crystals, particles are "trying" various sites of the growing object, until the most stable configuration is found. In this type of growth process a continuous rearrangement of particles takes place, the process is relatively slow, and the resulting objects are very regular (Sander 1987). Many growth processes in nature are not in equilibrium, aggregation of particles being an extreme example: as soon as a particle is added to the growing cluster, it stops trying other sites and no further rearrangement takes place. In this type of process the local chances that the object grows are not everywhere equal on the object and an unstable situation emerges. Typical for these phenomena is that they occur in a field which is in a steady state (compare the diffusion equation (2.7) when $\frac{dc}{dt}$ equals zero). The probability that growth takes place is the highest at the steepest gradients of the field, causing still steeper gradients. In Fig. 2.14 can be seen that the growing tips will press further into the rubber sheet, resulting in steeper gradients and a more instable situation. Growth processes in non-equilibrium are self-amplifying and relatively fast, and the resulting objects are often fractals.

Growth processes in which clusters of particles are formed and grow by adding particles to the cluster can be simulated with cellular automata. The particles are represented by sites in a lattice in these automata. Growth processes in equilibrium can be simulated with deterministic cellular automata (Wolfram 1983). For modelling non-equilibrium growth processes probabilistic cellular automata are more suitable. This method will also be applied in this book for modelling biological objects. For this reason the construction of a probabilistic cellular automaton in a steady state field will be discussed briefly.

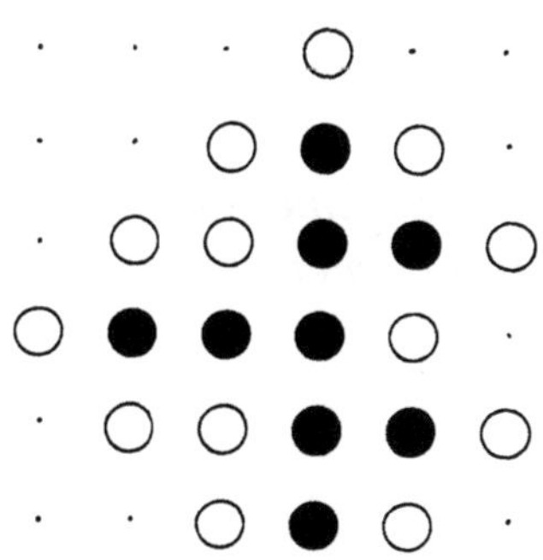

Fig. 2.16. Construction of a DLA cluster in a two dimensional lattice. The object itself consists of occupied sites in the lattice, which are displayed as black circles. The possible candidates which can be added to the object in a next growth step are indicated as open circles.

Laplacian growth can be simulated in a two-dimensional lattice, and the simulations can be extended to three or more dimensions (see Meakin 1983a and b). In the simulations shown in this section, growth starts with one occupied site in the lattice (the "seed" in Fig. 2.16). The cluster may look after a few initial growth steps as shown in Fig. 2.16. The occupied sites are displayed as black circles. In next growth steps new sites are added to the cluster; the possible candidates are indicated as white circles. The probability p that k, an element from the set of open circles $\circ$ neighbouring a black circle $\bullet$, will be added to the set of black circles is given by (2.8):

$$p(k \in \circ \rightarrow k \in \bullet) = \frac{(c_k)^\eta}{\sum_{j \in \circ}(c_j)^\eta} \qquad \text{where } c_k = \text{concentration at position } k \tag{2.8}$$

In (2.8) an exponent η is assumed to describe the relation between the local field and the probability (η usually ranges from 0.5 to 2.0, see Niemeyer et al. 1984; Meakin 1986). The sum in the denominator represents the sum of all local concentrations of the possible growth candidates (the open circles in Fig. 2.16).

Approximation solution Laplace equation

The concentrations in the lattice sites (lattice coordinates i, j), before each growth step, can be determined with the Laplace equation (2.7). The solution of this equation can be approximated by the following algorithm (see Ames 1977, Niemeyer et al. 1984, Press et al. 1988):

$$\textbf{while}\ (\ ((c_{i,j})_n - (c_{i,j})_{n-1}) > \textit{tolerance}\ )\{ \qquad (2.9)$$
$$c_{i,j} = \tfrac{1}{4}(c_{i+1,j} + c_{i-1,j} + c_{i,j+1} + c_{i,j-1})$$
$$\}$$

n = iteration number

This step in the simulation model, where the curved surface shown in Fig. 2.14 is determined, is computationally the most expensive part.

The local concentrations are visualized as alternating black and coloured regions in Fig. 2.15 (compare Mandelbrot and Evertsz 1990). The nutrient concentration decreases when the black or coloured basin of equal concentration range is situated closer to the object. The concentration decreases in the coloured basins when the colour shifts from pink to blue. The object itself is displayed in red, and the basin (with concentration near zero) where the object is located is coloured black. In this example a linear source (the top row of the lattice) of nutrient was chosen, for reasons which will become obvious later on. This figure shows an example of a DLA cluster on a 1000 x 1000 lattice, where a value of $\eta = 1$ was used in (2.8).

Fractal dimension DLA cluster

There are many more possibilities, like point-like or line-shaped nutrient sources. These boundary conditions do not affect the fractal dimensionalities of the generated structures (see Meakin 1986). The fractal dimension of the object shown in Fig. 2.15 is about 1.7 when the value $\eta = 1.0$ is used in (2.8). It is possible to generate objects with a higher fractal dimension by using a lower value for η. When a lower value for η is used the overall nutrient gradient around the object will become steeper. Although growth is fastest at the tips of the object, where the highest gradients occur, it can intuitively be seen that the probability that branches are formed at sites situated more in the bays of the object increases when such a lower value is used. The consequence will be that the overall branching degree increases and an object with a higher fractal dimension emerges.

The DLA model is typically suitable for modelling growth patterns in biology when the organisms can be considered as aggregates of loose

DLA model and bacteria colonies

particles. As a model of a growing coherent structure it is less applicable. An example of such a coherent structure is the formation of growth layers, in which neighbouring particles become connected with each other in a more systematic way. When the particles are connected in a mesh this has important consequences for the resulting growth form, as will be shown in the section on modelling radiate accretive growth. The DLA model has been applied in biology to model the forms of bacteria colonies (see Fujikawa and Matsushita 1989 and 1991; Matsushita and Fujkawa 1990; Matsuyama et al. 1989) and growth forms of dendritic hermatypic corals (see Nakamori 1988). One important feature of the DLA models in the present context is that the physical environment where the growth process takes place can be modelled and can be used to explain the emergence of growth forms. In general it is not easy and even quite artificial to describe the growth of an organism with a cellular automaton. For this reason, in a later section, the DLA model will be used in combination with a geometric model.

2.5 Generation of Fractal Objects Using Iterated Function Systems

Another method for the calculation and specification of objects, with a resemblance to biological objects, is based on Iterated Function Systems (IFS, see Barnsley 1988). With this method a large class of objects, which are often fractals, can be generated.

The first component of an IFS consists of a finite set of mappings of a 2- or 3-dimensional space into itself:

$$M = \{M_1, M_2, ..., M_m\}$$

The second component is a set of corresponding probabilities:

$$P = \{P_1, P_2, ..., P_m\}$$

in which:

$$\sum_{i=1}^{m} P_i = 1$$

In a number of cases, fractal objects may be generated by randomly choosing mappings from M. The iteration process starts with a point z_0, and a mapping M_i (with probability P_i) is chosen out of M, resulting in $z_1 = M_i(z_0)$. The result of the IFS is calculated in an iteration process (see Fig. 2.2). The objects X_n and X_{n+1} in the iteration process are, in this case, represented by 2- or 3-dimensional points and f by the set of mappings M with corresponding probabilities P. The initial points z_0 are driven by the iteration process to points of attraction ("the attractor"), in case the mappings are contractions and the process converges.

Dragon sweep

One example of the generation of a fractal object is the Dragon Sweep from Mandelbrot (1983). This object may be generated using two mappings in the complex plane (see Demko et al. 1985):

$$M_1 : z_{n+1} = sz_n + 1 \qquad (2.10)$$
$$M_2 : z_{n+1} = sz_n - 1$$

In these mappings z is a complex variable ($z = x + iy$) and s a complex parameter:

$$s = \frac{i}{2} + \frac{1}{2}$$

The corresponding set of probabilities is:

$$P = \{0.5, 0.5\}$$

Twin Dragon sweep

The process starts with the point 0 in the complex plane. The resulting figure, the Twin Dragon Sweep, is shown in Fig. 2.17. It can be demonstrated that this object is the same as the curve in Fig. 2.10B: when the latter is reflected, it fits exactly in the original one and together they form the Twin Dragon Sweep.

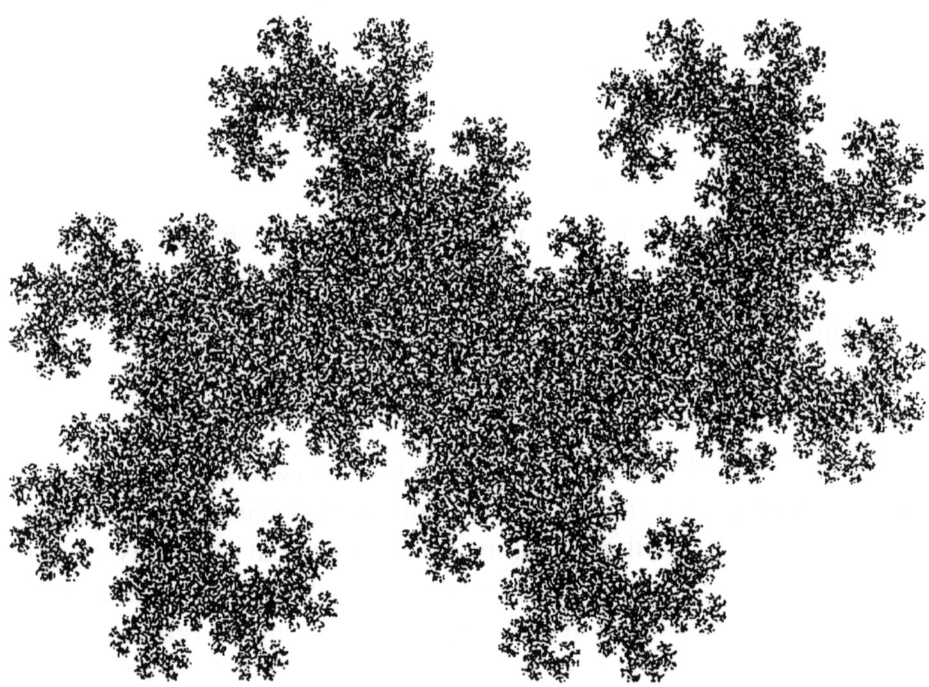

Fig. 2.17. Twin Dragon sweep generated with Iterated Function Systems, using the two mappings from (2.10)

A B

Fig. 2.18. Image of a leaf generated with Iterated Function Systems, using the four mappings from (2.11). In leaf A the set of corresponding probabilities from (2.11) was used. The choice of mappings is visualized in (B) in this picture the resulting points of the four mappings are separated from each other by translating M_1 .. M_4 respectively by the vectors: $0.1i$; 0; $0.1 + 0.1i$; $-0.1 + 0.1i$. Without applying these additional translations the four objects in B result in the object in A.

Most of the mappings in this section are displayed in complex notation, since this is usual in the literature on IFS. With this notation a compact description of the mappings is achieved. The complex mappings are only useful to describe the mappings in 2D. Alternatively, more in agreement with the next section, these mappings could be denoted as a concatenation of homogeneous transformations (scalings, rotations and translations).

The IFS method may be applied for generating images of natural objects, by searching for a "fitting" attractor. For this purpose it is necessary to cover the original object with locally affine images of itself. After having found the appropriate mappings (which is not always a trivial task) and selecting corresponding probabilities, an attractor is generated which approximates the original object. An example of the construction of an image of a leaf (see Barnsley 1988) by recursively applying a set of four mappings and a corresponding set of probabilities (2.11) is shown in Fig. 2.18A.

$$\begin{aligned}
M_1 : z_{n+1} &= 0.6z_n + (1 - 0.6)(0.45 + 0.9i) \qquad (2.11)\\
M_2 : z_{n+1} &= 0.6z_n + (1 - 0.6)(0.45 + 0.3i)\\
M_3 : z_{n+1} &= (0.4 - 0.3i)z_n + (1 - 0.4 + 0.3i)(0.60 + 0.9i)\\
M_4 : z_{n+1} &= (0.4 + 0.3i)z_n + (1 - 0.4 - 0.3i)(0.30 + 0.9i)\\
P1_{\text{leaf}} &= \{0.25, 0.25, 0.25, 0.25\}
\end{aligned}$$

Fig. 2.19. Branching structure resulting from the IFS in (2.12)

The choice of the mappings is visualized in Fig. 2.18B; in this picture the resulting points of the four mappings are separated from each other by translating M_1 .. M_4 respectively by the vectors: $0.1i$; 0; $0.1+0.1i$; $-0.1+0.1i$. Without applying these additional translations the four objects in B result in the object in A. In this figure it can be seen that the image of the leaf contains four smaller replicas of itself which can be generated by applying a combination of a translation, rotation and a scaling. After defining the four transformations delivering the replicas, the image can be described by the IFS in (2.11). A branching structure (see Lauwerier 1987) can be generated by using the following IFS:

IFS and branching objects

$$
\begin{aligned}
M_1 &: \begin{cases} x_{n+1} = 0.5x_n + 0.5y_n - 0.5 \\ y_{n+1} = 0.5x_n - 0.5y_n + 0.5 \end{cases} \qquad (2.12) \\
M_2 &: \begin{cases} x_{n+1} = 0.6667x_n + 0.3333 \\ y_{n+1} = -0.6667y_n \end{cases} \\
P1_{\text{branches}} &= \{0.5, 0.5\}
\end{aligned}
$$

The resulting branch is depicted in Fig. 2.19.

This method is suitable for approximating a given image and can be applied in data compression and for describing the self-similar aspects of the image. It is not suitable as a model of a growth process, as the mappings in (2.11), while generating the leaf image, do not supply any information about how the leaf was formed in a growth process.

2.6 Iterative Geometric Constructions

In the last method (see also Kaandorp 1987; Lauwerier and Kaandorp 1988) to be discussed, the objects are generated by geometric constructions. In geometric constructions the symbols X_n and X_{n+1} in Fig. 2.2 represent geometrical objects (edges, polylines, surfaces, volumes). In the iteration process, objects are replaced by further sets of objects. In many cases this process results in fractal objects. In the first subsection it is described how production rules can be formulated for an extensive class of objects. It will be demonstrated how from simple rules, stepwise, more complex rules can be built. Some of the fractal curves from Mandelbrot (1983) will be used as examples. From this starting point a robust system is developed, with which a large variety of objects can be generated. The objects used as examples do not have any biological significance and are only meant to demonstrate the development of the modelling system. In the second subsection it is shown how production rules can be entered into a 2D modelling system for iterative constructions. The final 2D geometric modelling system developed is suitable for simulating simple growth processes in 2D.

Geometric substitution

2.6.1 Geometric Production Rules in 2D Modelling

Many of the fractal curves from Mandelbrot (1983) can be constructed by defining an initial polyline (the *initiator*) and a generator polyline (the *generator*), which replaces the edges of the initiator in the iteration process. With those two components production rules for many objects, often characterized by a fractal dimension, may be formulated.

Koch curves

One of the quadric Koch curves (see object A in Fig. 2.20) results from the production rule in Fig. 2.21. The *initiator* and the *generator* are both represented by a list of edges, which contains all the geometric information.

In Fig. 2.20 the *initiator* component is indicated as *objects*: in the iteration process the *initiator* is the 0-approximant of a curve, which can have a fractal dimension. The third component in the production rule (the *base element*) represents the polyline being replaced in the iteration process. In the example of the quadric Koch curve this *base element* consists of one edge.

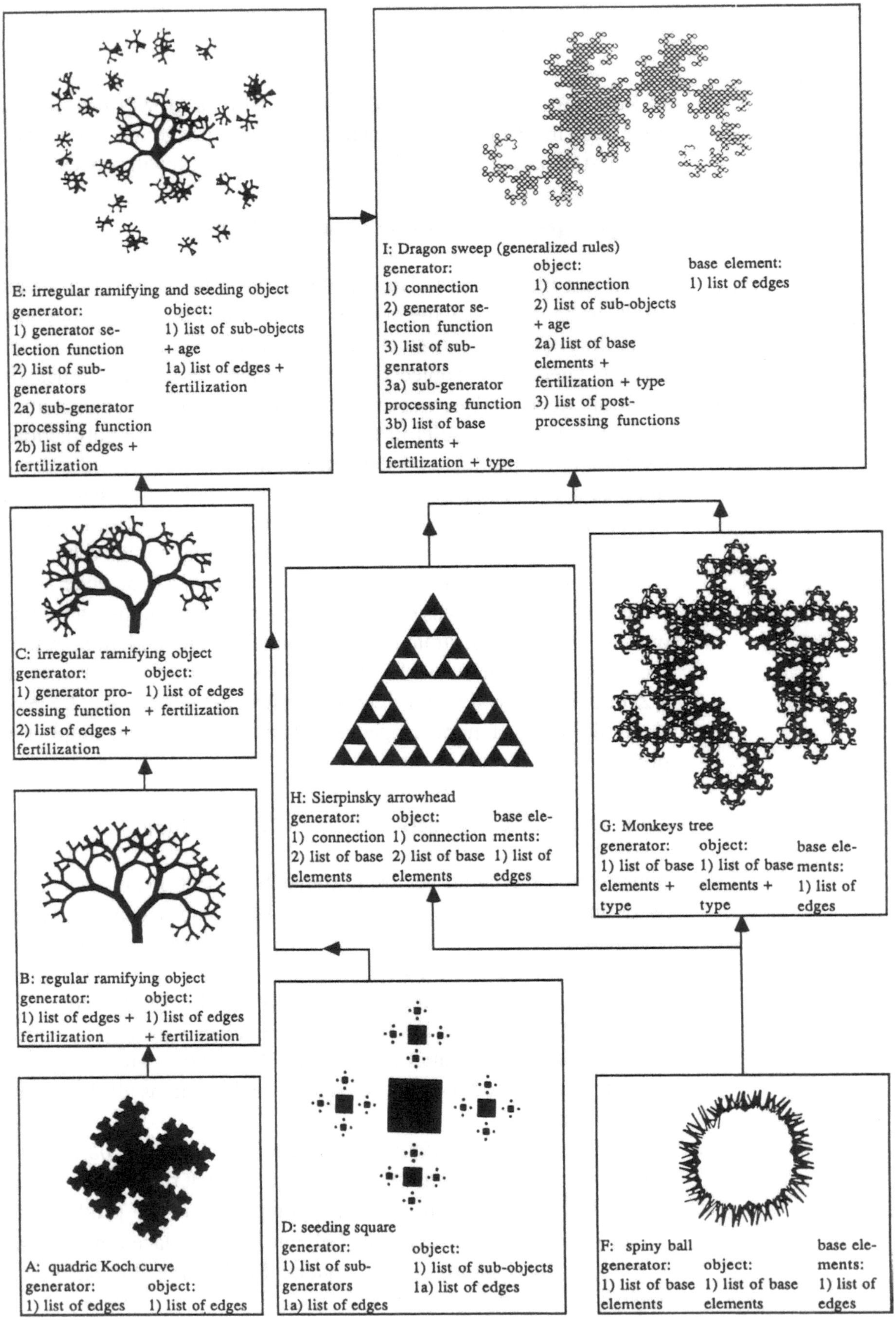
E: irregular ramifying and seeding object
generator:
1) generator selection function
2) list of sub-generators
2a) sub-generator processing function
2b) list of edges + fertilization
object:
1) list of sub-objects + age
1a) list of edges + fertilization
I: Dragon sweep (generalized rules)
generator:
1) connection
2) generator selection function
3) list of sub-genrators
3a) sub-generator processing function
3b) list of base elements + fertilization + type
object:
1) connection
2) list of sub-objects + age
2a) list of base elements + fertilization + type
3) list of post-processing functions
base element:
1) list of edges
C: irregular ramifying object
generator:
1) generator processing function
2) list of edges + fertilization
object:
1) list of edges + fertilization
H: Sierpinsky arrowhead
generator:
1) connection
2) list of base elements
object:
1) connection
2) list of base elements
base elements:
1) list of edges
G: Monkeys tree
generator:
1) list of base elements + type
object:
1) list of base elements + type
base elements:
1) list of edges
B: regular ramifying object
generator:
1) list of edges + fertilization
object:
1) list of edges + fertilization
A: quadric Koch curve
generator:
1) list of edges
object:
1) list of edges
D: seeding square
generator:
1) list of sub-generators
1a) list of edges
object:
1) list of sub-objects
1a) list of edges
F: spiny ball
generator:
1) list of base elements
object:
1) list of base elements
base elements:
1) list of edges

initiator	generator	base element
	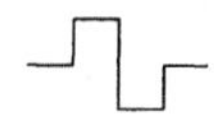	

Fig. 2.21. Production rule of a quadric Koch curve. The resulting fractal is shown in Fig. 2.20A.

In general the geometric construction, as described above, can be described as a *base element* (edge) replacement system:

$$\begin{aligned}
base\ element &= edge(V_a, V_b) \qquad (2.13)\\
initiator &= edge(V_0, V_1); edge(V_1, V_2)..edge(V_{n-1}, V_n);\\
generator &= edge(V_i, V_{i+1}); \rightarrow edge(V_i, M_{1j}(V_i));\\
&\quad edge(M_{1j}(V_i), M_{2j}(V_i)); \cdots edge(M_{m-1,j}(V_i), V_{i+1});\\
< \text{iteration} > & \quad < \text{iterated list of edges} >\\
0: & \quad \cdots edge(V_i, V_{i+1}); \cdots\\
1: & \quad \cdots edge(V_i, M_{11}(V_i)); edge(M_{11}(V_i), M_{21}(V_i)); \cdots\\
& \quad edge(M_{m-1,1}(V_i), V_{i+1}); \cdots\\
2: & \quad \cdots edge(V_i, M_{12}(V_i)); edge(M_{12}(V_i), M_{22}(V_i)); \cdots\\
& \quad edge(M_{m-1,2}(V_i), M_{11}(V_i));\\
& \quad edge(M_{11}(V_i), M_{12}(M_{11}(V_i)));\\
& \quad edge(M_{12}(M_{11}(V_i)), M_{22}(M_{11}(V_i))); \cdots\\
& \quad edge(M_{m-1,2}(M_{11}(V_i)), M_{21}(V_i));\\
& \quad edge(M_{21}(V_i), M_{12}(M_{21}(V_i)));\\
& \quad edge(M_{12}(M_{21}(V_i)), M_{22}(M_{21}(V_i))); \cdots\\
& \quad edge(M_{m-1,2}(M_{21}(V_i)), M_{21}(V_i)); \cdots\\
& \quad edge(M_{m-1,1}(V_i), M_{12}(M_{m-1,1}(V_i)));\\
& \quad edge(M_{12}(M_{m-1,1}(V_i)), M_{22}(M_{m-1,1}(V_i)));\\
& \quad edge(M_{m-1,2}(M_{m-1,1}(V_i)), V_{i+1}); \cdots
\end{aligned}$$

Fig. 2.20. Classification diagram of linear fractals based on the minimal rules necessary for representing all components of the production rules. The general set of rules on top of the classification (I) can be used for representing all objects discussed in Sect. 2.6.

In this edge replacement system the iteration process starts with an initial polyline consisting of $n + 1$ vertices V and $n + 1$ edges. In the generator an $edge(V_i, V_{i+1})$ $(0 \leq i \leq n)$ is replaced by a series of m new edges. The new edges are obtained by a series of transformations, derived from the geometrical information described in the *generator* polyline (see Fig. 2.21). Figure 2.22 shows how the series of $m - 1$ transformations $M_{1,j}..M_{m-1,j}$ are obtained from the *generator* polyline.

The transformations are determined in four steps:

1) In the first step the start points of the *generator* polyline (VG_0) and the edge being replaced (V_i) are positioned so as to coincide.
2) In the second step the angle γ between the $edge(V_i, V_{i+1})$ and $edge(VG_0^*, VG_m^*)$ is determined, where as rotation point the vertex V_i is used.
3) In the third step the scaling factor:
$$sf = \frac{\|V_i, V_{i+1}\|}{\|VG_0, VG_m\|} \tag{2.14}$$
is determined.
4) In the fourth step a translation is done over the vector $[V_{xk} - V_{xi}, V_{yk} - V_{yi}]$ $(0 \leq k \leq m)$.

In each step a combination of a scaling S, rotation R and translation T is done for each edge. Together this combination can be written as a linear transformation:

$$M_{i,j} = R_{V_i}(\gamma).S(sf, sf).T(V_{xk} - V_{xi}, V_{yk} - V_{yi}) \tag{2.15}$$
$$\text{for } 0 \leq k \leq m,$$
$$\text{where } j \text{ is iteration number}$$

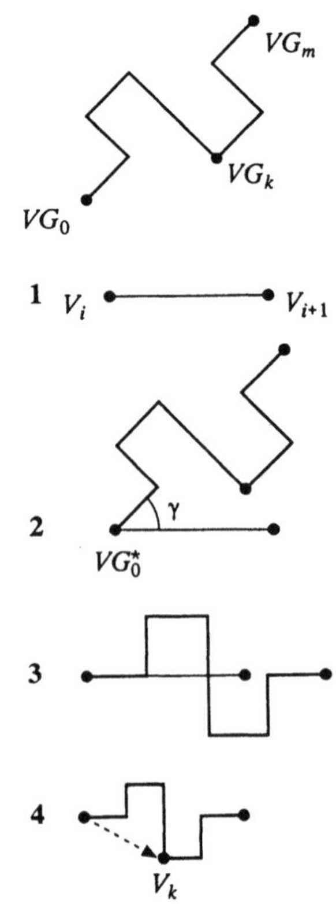

Fig. 2.22. Derivation of the $m - 1$ transformations $M_{1j}..M_{m-1,j}$ from the *generator* polyline. As example the *generator* from Fig. 2.21 is used.

In this respect the Koch curve and the other fractals which will be discussed can be described as linear fractals.

The linear fractals can be classified on the basis of the set of rules minimally necessary to represent all components of the production rule. In Fig. 2.20 a possible classification is shown of some linear fractals. The quadric Koch curve (object A) is represented by the smallest number of rules and is the most primitive object. The rules in this diagram are the specifications of the actual data structures, which were used in the implementation of the geometric modelling system.

Fertilization attribute

An attribute can be introduced in the production rule which defines the *fertilization* state of each edge. This *fertilization* attribute may be in the state "fertile" or "not-fertile". This attribute controls which edges will be replaced in the next iteration steps; only the fertile edges of the preceding iteration step are replaced by the *generator*. The production rule for a Pythagoras tree (Fig. 2.23) is an example. In this rule the *fertilization* attributes of the edges of the *initiator* and *generator* are defined; in Fig. 2.23 the fertile edges are indicated with asterisks. In Fig. 2.20B the list of edges representing the ramiform object is extended with a *fertilization* attribute. The edge replacement system for the tree construction can be described as:

Pythagoras tree

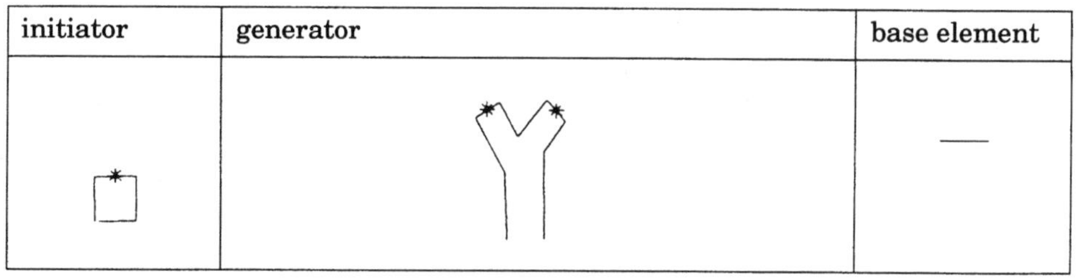

Fig. 2.23. Production rule for a simple tree (Pythagoras tree). Fertile edges (edges of the preceding iteration step which will be replaced in the next iteration step) are marked with asterisks. The resulting object is shown in Fig. 2.20B.

$$
\begin{aligned}
\textit{base element} &= edge(V_a, V_b) \qquad (2.16)\\
\textit{initiator} &= (edge(V_0, V_1), NF); (edge(V_1, V_2), F);\\
&\quad (edge(V_2, V_3), NF); (edge(V_3, V_0), NF);\\
\textit{generator} &= (edge(V_i, V_{i+1}), F) \rightarrow (edge(V_i, M_{1j}(V_i)), NF);\\
&\quad (edge(M_{1j}(V_i), M_{2j}(V_i)), NF);\\
&\quad (edge(M_{2j}(V_i), M_{3j}(V_i)), F);\\
&\quad (edge(M_{3j}(V_i), M_{4j}(V_i)), NF);\\
&\quad (edge(M_{4j}(V_i), M_{5j}(V_i)), NF);\\
&\quad (edge(M_{5j}(V_i), M_{6j}(V_i)), F);\\
&\quad (edge(M_{6j}(V_i), M_{7j}(V_i)), NF);\\
&\quad (edge(M_{7j}(V_i), V_{i+1}), NF);\\
&\quad (edge(V_i, V_{i+1}), NF); \rightarrow (edge(V_i, V_{i+1}), NF);
\end{aligned}
$$

< iteration > < iterated list of edges >

$$
\begin{aligned}
0: &\quad (edge(V_0, V_1), NF); (edge(V_1, V_2), F);\\
&\quad (edge(V_2, V_3), NF); (edge(V_3, V_0), NF);\\
1: &\quad (edge(V_0, V_1), NF); (edge(V_1, M_{11}(V_1)), NF);\\
&\quad (edge(M_{11}(V_1), M_{21}(V_1)), NF);\\
&\quad (edge(M_{21}(V_1), M_{31}(V_1)), F);\\
&\quad (edge(M_{31}(V_1), M_{41}(V_1)), NF);\\
&\quad (edge(M_{41}(V_1), M_{51}(V_1)), NF);\\
&\quad (edge(M_{51}(V_i), M_{61}(V_1)), F);\\
&\quad (edge(M_{61}(V_1), M_{71}(V_1)), NF);\\
&\quad (edge(M_{71}(V_1), V_2), NF); (edge(V_2, V_3), NF);\\
&\quad (edge(V_3, V_0), NF);
\end{aligned}
$$

In this replacement system the *fertilization* attribute is indicated as F (fertile) or NF (not-fertile).

The next step, in creating an extensive class of production rules, is to create a continuous range of *generators*. This was done by introducing a new attribute in the description of the *generator*: the *generator processing function*. In the new construction the original *generator* is processed by this function. The new *generator* is a transformation of the original one and may vary between certain limits, as specified by the *generator processing function*. An example is shown in Fig. 2.24, where a *generator processing function* is introduced in the construction of a ramiform object. In this example a function is introduced which processes the original generator by rotating it between two limits. The angle of rotation θ is determined by a random function and an irregular ramifying object is obtained (see Fig. 2.20C). The construction of the irregular ramifying object is an example of a non-deterministic iterative geometric construction and can be compared with the L-system where randomness was applied (Fig. 2.13). In the classification diagram (Fig. 2.20C) it can be seen that the *generator* component is extended by a new attribute, which contains a reference to a *generator processing function*. The edge replacement system for this construction can be described as:

Generator processing function

Non-deterministic iterative geometric constructions

$$base\ element = edge(V_a, V_b) \tag{2.17}$$

$$initiator = (edge(V_0, V_1), NF); (edge(V_1, V_2), F); (edge(V_2, V_3), NF); (edge(V_3, V_0), NF);$$

generator processing function = for each $(edge(V_i, V_{i+1}), F)$ a rotation matrix R_θ is determined for $lower\_limit \leq \theta \leq upper\_limit$; θ is chosen from a uniform distribution with two limits

$$\begin{aligned} generator = {} & (edge(V_i, V_{i+1}), F); \rightarrow \\ & (edge(V_i, R_\theta(M_{1j}(V_i)), NF); \\ & (edge(R_\theta(M_{1j}(V_i)), R_\theta(M_{2j}(V_i))), NF); \\ & (edge(R_\theta(M_{2j}(V_i)), R_\theta(M_{3j}(V_i))), F); \\ & (edge(R_\theta(M_{3j}(V_i)), R_\theta(M_{4j}(V_i))), NF); \\ & (edge(R_\theta(M_{4j}(V_i)), R_\theta(M_{5j}(V_i))), NF); \\ & (edge(R_\theta(M_{5j}(V_i)), R_\theta(M_{6j}(V_i))), F); \\ & (edge(R_\theta(M_{6j}(V_i)), R_\theta(M_{7j}(V_i))), NF); \end{aligned}$$

initiator	generator	base element

Fig. 2.24. Production rule for a tree in which the original *generator* is processed by a function which allows random movements of the *generator* between two limits. The *generator processing function* is described in the right part of the *generator* component. The resulting object is shown in Fig. 2.20C.

initiator	generator	base element
	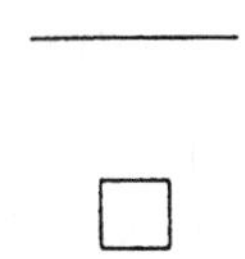	

Fig. 2.25. Production rule for a self-seeding square. The *generator* consists of two parts and seeds new squares during each iteration. The resulting object is shown in Fig. 2.20D.

$$(edge(R_\theta(M_{7j}(V_i)), V_{i+1}), NF);$$
$$(edge(V_i, V_{i+1}), NF); \rightarrow$$
$$(edge(V_i, V_{i+1}), NF);$$

In this system, for each fertile edge the transformation M_{ij} is extended with a rotation over the angle θ.

Self-seeding square

Another class of production rules contains rules which generate new objects in each iteration step. The result of such a rule is a fragmented cluster of objects. Each enlargement of the fragmented curve reveals that the object is still further subdivided into sub-objects. An example of the construction of such an object, which generates recursively new objects, is displayed in the production rule in Fig. 2.25, the result of which is displayed in Fig. 2.20D. In each iteration step each edge generates an new object, a square. In order to represent this type of objects, both the *generator* and the *initiator* component are extended, the *generator* by a list of *sub-generators* and the *initiator* by a list of *objects*.

Each *sub-generator* and *object* consists of a list of edges. This construction is described in the following replacement system:

$$\begin{aligned}
initiator &= edge(V_0, V_1); edge(V_1, V_2); edge(V_2, V_3); \quad (2.18)\\
&\quad edge(V_3, V_0);\\
generator &= edge(V_i, V_{i+1}); \rightarrow\\
&\quad sub_generator0:\\
&\quad edge(V_i, V_{i+1});\\
&\quad sub_generator1:\\
&\quad edge(M_{1j}(V_i), M_{2j}(V_i)); edge(M_{2j}(V_i), M_{3j}(V_i));\\
&\quad edge(M_{3j}(V_i), M_{4j}(V_i)); edge(M_{4j}(V_i), M_{1j}(V_i));
\end{aligned}$$

< iteration > < iterated list of edges >

$$\begin{aligned}
0: &\quad object0:\\
&\quad edge(V_0, V_1); edge(V_1, V_2);\\
&\quad edge(V_2, V_3); edge(V_3, V_0);\\
1: &\quad object0:\\
&\quad edge(V_0, V_1); edge(V_1, V_2);\\
&\quad edge(V_2, V_3); edge(V_3, V_0);\\
&\quad object1:\\
&\quad edge(M_{11}(V_0), M_{21}(V_0)); edge(M_{21}(V_0), M_{31}(V_0));\\
&\quad edge(M_{31}(V_0), M_{41}(V_0)); edge(M_{41}(V_0), M_{11}(V_0));\\
&\quad object2:\\
&\quad edge(M_{11}(V_1), M_{21}(V_1)); edge(M_{21}(V_1), M_{31}(V_1));\\
&\quad edge(M_{31}(V_1), M_{41}(V_1)); edge(M_{41}(V_1), M_{11}(V_1));\\
&\quad object3:\\
&\quad edge(M_{11}(V_2), M_{21}(V_2)); edge(M_{21}(V_2), M_{31}(V_2));\\
&\quad edge(M_{31}(V_2), M_{41}(V_2)); edge(M_{41}(V_2), M_{11}(V_2));\\
&\quad object4:\\
&\quad edge(M_{11}(V_3), M_{21}(V_3)); edge(M_{21}(V_3), M_{31}(V_3));\\
&\quad edge(M_{31}(V_3), M_{41}(V_3)); edge(M_{41}(V_3), M_{11}(V_3));
\end{aligned}$$

Generator selection function

A new class of objects is created by applying several types of generators in the iteration process. The *generator* component in the production rule is represented in this class by a list of *generators* and a *generator selection function*, which defines which *generator* should be applied to a certain edge. An example of a rule for an irregular ramifying object, which generates new irregular ramifying objects, is shown in Fig. 2.26. In this example a list of *generators* consists of two elements: the irregular ramifying *generator* (Fig. 2.24) and the *generator* which "seeds" squares

Fig. 2.26. Production rule for a vegetation of ramifying objects. The growing *generator* (gen0) is chosen by the *generator selection function* (left part of the *generator* component) as long as the *age* of the *object* is less than 3 iterations; on the third iteration the *object* starts seeding (gen1). The resulting object is shown in Fig. 2.20E and 2.27.

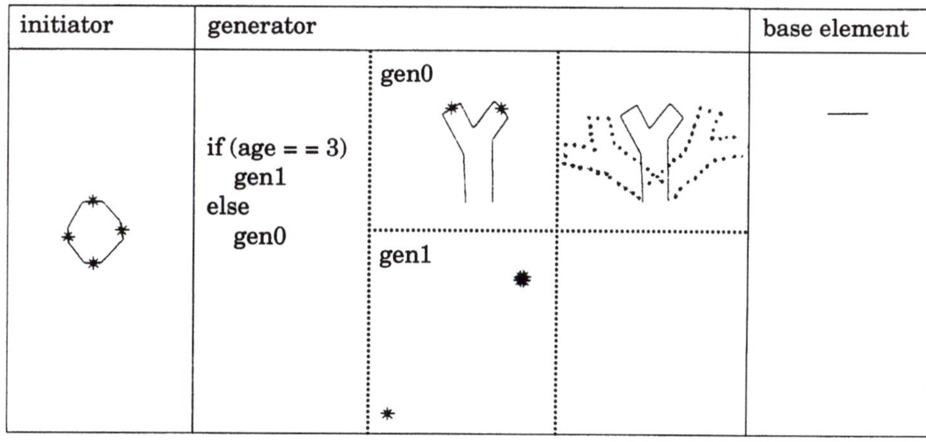

(Fig. 2.25). A *generator selection function* is included in the *generator* component and is used to determine which *generator* should be used for replacing edges of the preceding iteration step. The growing *generator* (*gen*0) is chosen as long as the *age* attribute of the *object* does not equals 3 iteration steps. When the *age* of the *object* equals three iteration steps, the *object* starts seeding a new set of *objects* (*age* equals 0), using the seeding *generator* (*gen*1). The result, a self-reproducing ramiform object is shown in Figs. 2.20E and 2.27. The original ramiform object has changed into a vegetation of ramiform objects. The construction is described in the following edge replacement system:

Vegetation of ramiform objects

Fig. 2.27. Vegetation of irregular ramifying objects produced with the rule in Fig. 2.26. In this picture three successive generations of ramifying objects are displayed.

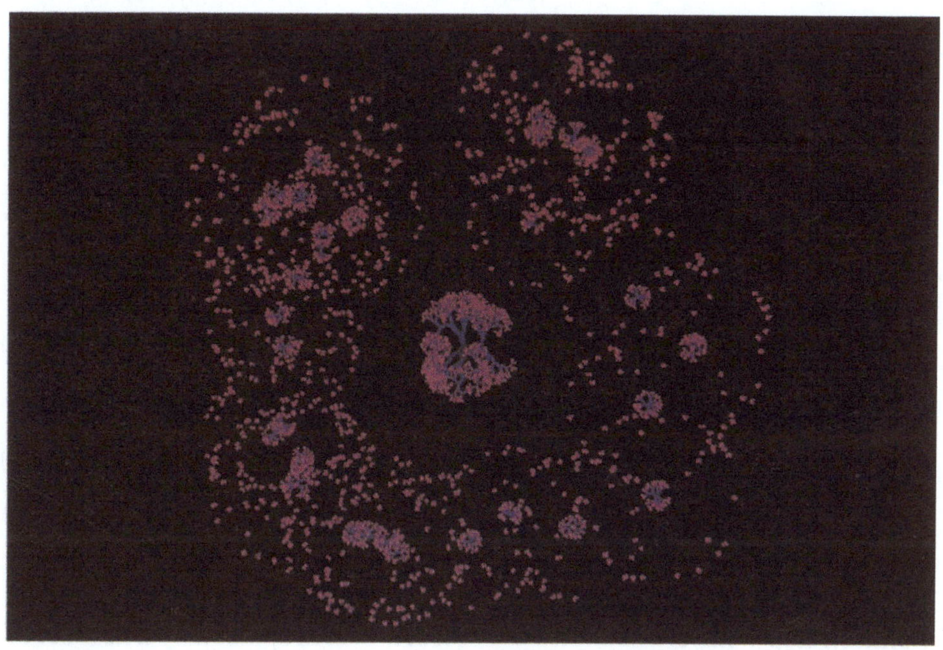

$$
\begin{aligned}
\textit{base element} &= edge(V_a, V_b) \qquad (2.19)\\
\textit{initiator} &= ((edge(V_0, V_1), F); (edge(V_1, V_2), NF);\\
&\quad (edge(V_2, V_3), F); (edge(V_3, V_4), NF);\\
&\quad (edge(V_4, V_5), F); (edge(V_5, V_6), NF);\\
&\quad (edge(V_6, V_7), F); (edge(V_7, V_0), NF);), age = 0);\\
\textit{generator processing}\\
\textit{function} &= \text{for each } (edge(V_i, V_{i+1}), F) \text{ a rotation}\\
&\quad \text{matrix } R_\theta \text{ is determined}\\
&\quad \text{for } lower\_limit \le \theta \le upper\_limit;\\
&\quad \theta \text{ is chosen from a uniform distribution}\\
&\quad \text{with two limits}\\
\textit{generator selection}\\
\textit{function} &= \text{if } (age == 3) \text{ then } generator \rightarrow generator1\\
&\quad \text{else } generator \rightarrow generator0\\
generator0 &= (edge(V_i, V_{i+1}), F); \rightarrow\\
&\quad (edge(V_i, R_\theta(M_{1j}(V_i)), NF);\\
&\quad (edge(R_\theta(M_{1j}(V_i)), R_\theta(M_{2j}(V_i))), NF);\\
&\quad (edge(R_\theta(M_{2j}(V_i)), R_\theta(M_{3j}(V_i))), F);\\
&\quad (edge(R_\theta(M_{3j}(V_i)), R_\theta(M_{4j}(V_i))), NF);\\
&\quad (edge(R_\theta(M_{4j}(V_i)), R_\theta(M_{5j}(V_i))), NF);\\
&\quad (edge(R_\theta(M_{5j}(V_i)), R_\theta(M_{6j}(V_i))), F);\\
&\quad (edge(R_\theta(M_{6j}(V_i)), R_\theta(M_{7j}(V_i))), NF);\\
&\quad (edge(R_\theta(M_{7j}(V_i)), V_{i+1}), NF);\\
generator1 &= (edge(V_i, V_{i+1}), F); \rightarrow\\
&\quad sub\_generator1, 0:\\
&\quad (edge(V_i, V_{i+1}), F);\\
&\quad sub\_generator1, 1:\\
&\quad ((edge(M_{1j}(V_i), M_{2j}(V_i)), F);\\
&\quad (edge(M_{2j}(V_i), M_{3j}(V_i)), F);\\
&\quad (edge(M_{3j}(V_i), M_{4j}(V_i)), F);\\
&\quad (edge(M_{4j}(V_i), M_{1j}(V_i)), F);), age = 0);\\
&\quad (edge(V_i, V_{i+1}), NF) \rightarrow (edge(V_i, V_{i+1}), NF);
\end{aligned}
$$

In all objects discussed so far, *base elements* (edges) of the preceding iteration step with *fertilization* state "fertile" were replaced by a new set of

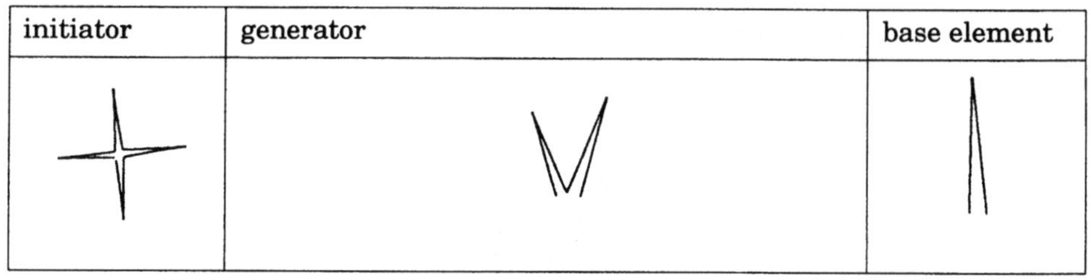

Fig. 2.28. Production rule for a simple *base element object.* The resulting object is shown in Fig. 2.20F.

Base element

edges. This rule can be generalized to a new rule where the *base element* consists of a polyline, which is replaced by a new set of polylines. The *base element*, the third component in the production rule, can consists of more than one edge. The rules in the classification diagram (Fig. 2.20F) are extended with this third component. An example of a rule where this extension is used is shown in Fig. 2.28. Here spine-like polylines, consisting of two edges, are replaced by a new set of spines. The result is a spiny ball.

In the *base element* replacement systems below, the notation $be(V_0, V_b)$ indicates a list of $b + 1$ edges:

$$be(V_0, V_b) = \mathcal{L}_{i=0}^{b}\{edge(V_i, V_{i+1})\} \tag{2.20}$$

The edges in the replacement systems to be illustrated, are connected, for $b > 1$ $be(V_0, V_b)$ can be written as a series of edges:

$$be(V_0, V_b) = be(edge(V_0, V_1); edge(V_1, V_2); \cdots edge(V_b, V_{b+1})); \tag{2.21}$$

Spiny ball

The construction of the spiny ball can be represented by the following *base element* replacement system:

$$\begin{aligned}
base\ element &= \mathcal{L}_{i=0}^{1}\{edge(V_i, V_{i+1}\} \qquad (2.22)\\
initiator &= be(V_0, V_2); be(V_2, V_4); be(V_4, V_6); be(V_6, V_0);\\
generator &= be(V_{2*i}, V_{2*i+2}); \rightarrow\\
&\quad be(V_{2*i}, M_{2j}(V_{2*i})); be(M_{2j}(V_{2*i}), V_{2*i+2});\\
<\text{iteration}> &\quad <\text{iterated list of base elements}>\\
0: &\quad be(V_0, V_2); be(V_2, V_4); be(V_4, V_6); be(V_6, V_0);\\
1: &\quad be(V_0, M_{21}(V_0)); be(M_{21}(V_0), V_2);\\
&\quad be(V_2, M_{21}(V_2)); be(M_{21}(V_2), V_4);\\
&\quad be(V_4, M_{21}(V_4)); be(M_{21}(V_4), V_6);\\
&\quad be(V_6, M_{21}(V_6)); be(M_{21}(V_6), V_0);
\end{aligned}$$

In this system the *base elements* are denoted as $be(\cdots)$.

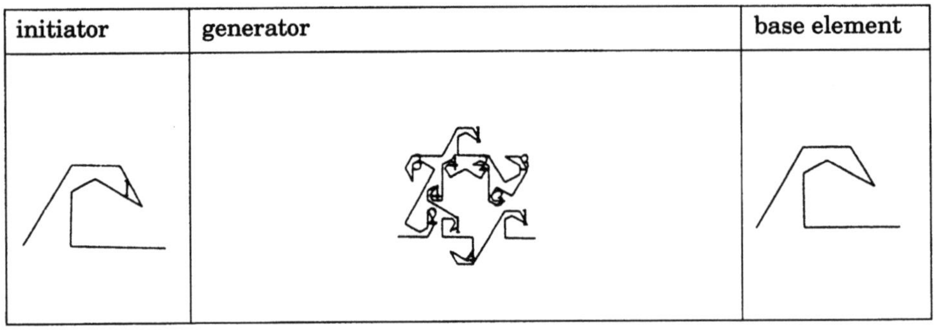

Fig. 2.29. Production rule for the Monkeys tree shown in Fig. 2.20G. The *type* of the *base elements* is denoted in the first two components with numbers: *type* 1, normal *base element*; *type* 2, y-coordinate has been reflected; *type* 3, x-coordinate has been reflected; *type* 4, x- and y-coordinate has been reflected.

Type attribute

Monkeys tree

A further generalization of the rule in which *base elements* are replaced by a new set of *base elements* is a rule where different versions of one type of *base element* are used. The different versions of the *base element* are obtained by reflecting the original *base element*. In this new generalization the attribute *type* is added to each *base element*, which indicates the type of reflection used. An example of a construction of an object with four *types* of reflection is displayed in Fig. 2.29. In this example the *types* are: *type* 1, normal base element; *type* 2, y-coordinate of base element is reflected; *type* 3, x-coordinate of base element is reflected; *type* 4, x- and y-coordinate of base element are reflected. As a result it is possible to construct, for example, the Monkeys tree of Mandelbrot (1983) (see diagram Fig. 2.20G). The construction of the Monkeys tree can be represented by the following *base element* replacement system:

$$\begin{aligned}
base\ element &= \mathcal{L}_{i=0}^{6}\{edge(V_i, V_{i+1}\} \qquad (2.23)\\
initiator &= (be(V_0, V_7), type1);\\
generator &= (be(V_{7*i}, V_{7*i+7}), type1); \rightarrow\\
&\quad (be(V_{7*i}, M_{7,1}(V_{7*i})), type2);\\
&\quad (be(M_{7,1}(V_{7*i}), M_{14,1}(V_{7*i})), type3);\\
&\quad (be(M_{14,1}(V_{7*i}), M_{21,1}(V_{7*i})), type1);\\
&\quad (be(M_{21,1}(V_{7*i}), M_{28,1}(V_{7*i})), type3);\\
&\quad (be(M_{28,1}(V_{7*i}), M_{35,1}(V_{7*i})), type3);\\
&\quad (be(M_{35,1}(V_{7*i}), M_{42,1}(V_{7*i})), type4);\\
&\quad (be(M_{42,1}(V_{7*i}), M_{49,1}(V_{7*i})), type4);\\
&\quad (be(M_{49,1}(V_{7*i}), M_{56,1}(V_{7*i})), type1);\\
&\quad (be(M_{56,1}(V_{7*i}), M_{63,1}(V_{7*i})), type4);\\
&\quad (be(M_{63,1}(V_{7*i}), V_{7*i+7}))), type1);\\
&\quad (be(V_{7*i}, V_{7*i+7}), type2); \rightarrow \cdots\\
&\quad (be(V_{7*i}, V_{7*i+7}), type3); \rightarrow \cdots\\
&\quad (be(V_{7*i}, V_{7*i+7}), type4); \rightarrow \cdots
\end{aligned}$$

Types of base elements

In this system only the replacement of a *base element* of *type* 1 is displayed; the resulting *types* of *base elements* of the other types in the replacement process can be found in Table 2.1.

Table 2.1. Resulting *type* of *base element*, when a *base element* of a preceding iteration step is replaced by a new set of *base elements*

type base element being replaced	1	2	3	4
type 1 *generator*	1	2	3	4
type 2 *generator*	2	1	4	3
type 3 *generator*	3	4	1	2
type 4 *generator*	4	3	2	1

In the previous examples the polylines are connected in one or more curves; after each iteration step, in most cases, the length of each curve increases (only the length of the self-seeding square curves in Fig. 2.20D remains the same) but all elements are connected and there is one start and one end point of the curve. A more extended type of construction applies a *generator* and an *object*, which consists of *base elements*, which may be either connected or disconnected. This extension is introduced in the rules shown in Fig. 2.20H by adding the attribute *connection*, which can be in the state "connected" or "disconnected". An example is shown in Fig. 2.30.

Sierpinski arrowhead

This production rule specifies the construction of a Sierpinski arrowhead, using triangles as *base elements*. The *base element* replacement system is given below:

$$
\begin{aligned}
base\ element &= be(edge(V_a, V_b); edge(V_b, V_c); edge(V_c, V_a)) \\
initiator &= be(edge(V_0, V_1); edge(V_1, V_2); edge(V_2, V_0)); \\
generator &= be(edge(V_{3*i}, V_{2*i+1}); edge(V_{3*i+1}, V_{3*i+2})); \\
&\quad edge(V_{3*i+2}, V_{3*i})); \rightarrow \\
&\quad be(edge(V_{3*i}, M_{1j}(V_{3*i})); \\
&\quad edge(M_{1j}(V_{3*i}), M_{2j}(V_{3*i})); \\
&\quad edge(M_{2j}(V_{3*i}), V_{3*i})); \\
&\quad be(edge(M_{1j}(V_{3*i}), V_{3*i+1}); \\
&\quad edge(V_{3*i+1}), M_{3j}(V_{3*i})); \\
&\quad edge(M_{3j}(V_{3*i}), M_{1j}(V_{3*i+1})); \\
&\quad be(edge(M_{2j}(V_{3*i}), M_{3j}(V_{3*i})); \\
&\quad edge(M_{3j}(V_{3*i}), (V_{3*i+2})); \\
&\quad edge(V_{3+i+2}, M_{2j}(V_{3*i})));
\end{aligned}
\qquad (2.24)
$$

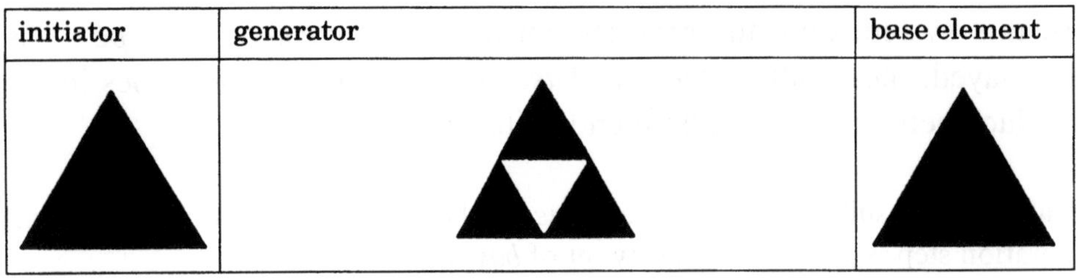

Fig. 2.30. Production rule for constructing a Sierpinski arrowhead. The resulting object is shown in Fig. 2.20H.

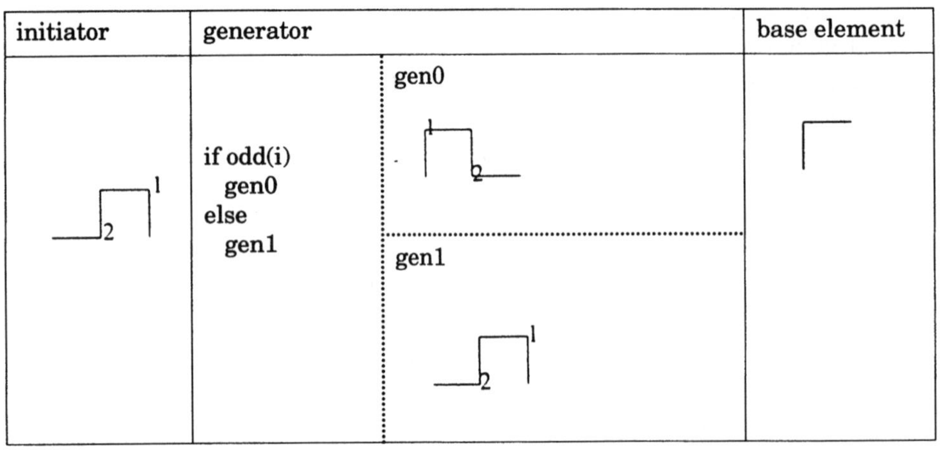

Fig. 2.31. Production rule for constructing the Dragon Sweep: odd *base elements* of the preceding object are replaced by *gen*0, even *base elements* by *gen*1. The resulting object is shown in Fig. 2.20I.

Dragon sweep

The final step in creating a general system of rules for the generation of objects is to combine all previous rules. This combination of rules is displayed in Fig. 2.20I: with this combination it is possible to capture all the objects shown so far and it is possible to define production rules for still more objects. An example of a new construction, which is possible with (a subset of) the rules in Fig. 2.20I, is the production rule (Fig. 2.31) of the Dragon sweep (Mandelbrot 1983). The Dragon sweep is created by using a list of two *generators* which alternate. The *base element* used in the *generator* and *object* component consists of two edges. The odd *base elements* of the preceding object are replaced by *gen*0, the even *base elements* by *gen*1. The construction can be described, using a subset of the replacement rules, in the following system:

$$\begin{aligned} base\ element &= \mathcal{L}_{i=0}^{1}\{edge(V_i, V_{i+1}\} \qquad (2.25) \\ initiator &= (be(V_0, V_2), type2); (be(V_2, V_4), type1); \\ generator\ selection & \\ function &= \text{if } (odd(i)) \text{ then } generator \rightarrow generator0 \\ & \quad \text{else } generator \rightarrow generator1 \end{aligned}$$

$$\begin{aligned}
generator0 &= (be(V_{2i}, V_{2*i+2}), type1); \rightarrow \\
&\quad (be(V_{2i}, M_2j(V_{2i})), type1); \\
&\quad (be(M_2j(V_{2i}), V_{2*i+2}), type2); \\
&\quad (be(V_{2i}, V_{2*i+2}), type2); \rightarrow \\
&\quad (be(V_{2i}, M_2j(V_{2i})), type2); \\
&\quad (be(M_2j(V_{2i}), V_{2*i+2})), type1); \\
generator1 &= (be(V_{2i}, V_{2*i+2}), type1); \rightarrow \\
&\quad (be(V_{2i}, M_5j(V_{2i})), type2); \\
&\quad (be(M_5j(V_{2i}), V_{2*i+2})), type1); \\
&\quad (be(V_{2i}, V_{2*i+2})), type2); \rightarrow \\
&\quad (be(V_{2i}, M_5j(V_{2i})), type1); \\
&\quad (be(M_5j(V_{2i}), V_{2*i+2})), type2);
\end{aligned}$$

In the previous sections it was demonstrated that this object can be constructed in many alternative ways: by applying formal languages (see Fig. 2.10B) or the Iterated Function Systems (see Fig. 2.17). It is an object which occurs frequently in the literature on iteration processes. Some authors even associate it with a pathological demon in science (see Lovelock 1988)!

Fig.2.20I shows that another rule (list of *post-processing functions*) was added, which was not used until then. This rule contains references to functions which post-process the result X_{n+1} (see Fig. 2.2) of an iteration step. These functions will be discussed in more detail in Sect. 2.6.3, where the iterative geometric constructions will be used to model a growth process.

2.6.2 The Geometric Modelling System for 2D Objects

The three components of the production rule may be regarded as arguments for an algorithm for calculating objects. The object is generated by an algorithm in an iteration process, in which the calculated object (*new_objects*) is used as input (*old_objects*) of a next iteration (compare X_{n+1} and X_n in Fig. 2.2). The algorithm, suitable for 2D objects, is described in a summarized form using pseudo code below:

```
fractal( old_objects, new_objects, generator, base_element ) {
   A: if ((no generator processing function is used) &&
   (no selection function is used))
      a local copy of the generator is made and the transformation
      matrices M_ij (see 2.15) are derived from the generator;
   B: next object from old_objects is taken {
      C1: next base_element from current old_object is taken {
         D1: if (current old base_element is fertile) {
            E1: if ((generator processing function is used) ||
            (selection function is used))
               In case a selection function is used a generator is
               selected from the list of generators;
               In case a generator processing function is used a local
               (copy changed by a generator processing function ) is
               made of the original generator, the transformation
               matrices M_ij are derived from the local copy;
            E2: next sub-generator is taken {
               F: next base_element is taken from the sub-generator {
                  G: next vertex is taken from base_element of the sub-generator {
                     G1: transformation is performed using M_ij;
                     new vertex is added to new_objects;
                  }
                  fertilization status of new base_element (equal to
                  fertilization status of current base_element generator)
                  is added to new_objects,
                  type of the new base_element is evaluated from type
                  current base_element generator and current old
                  base_element and added to new_objects (see Table 2.1);
               } end F
            } end E2
            E3: if (base_element generates new object)
               A new object is added to new_objects;
         } end D1
         D2: if (current old base_element is not fertile)
            old base_element and its fertilization and type state is added to new_objects;
      } end C1
      C2: if (post-processing functions are used)
         new_object is post-processed by one or more post-processing functions;
   } end B
} end fractal
```

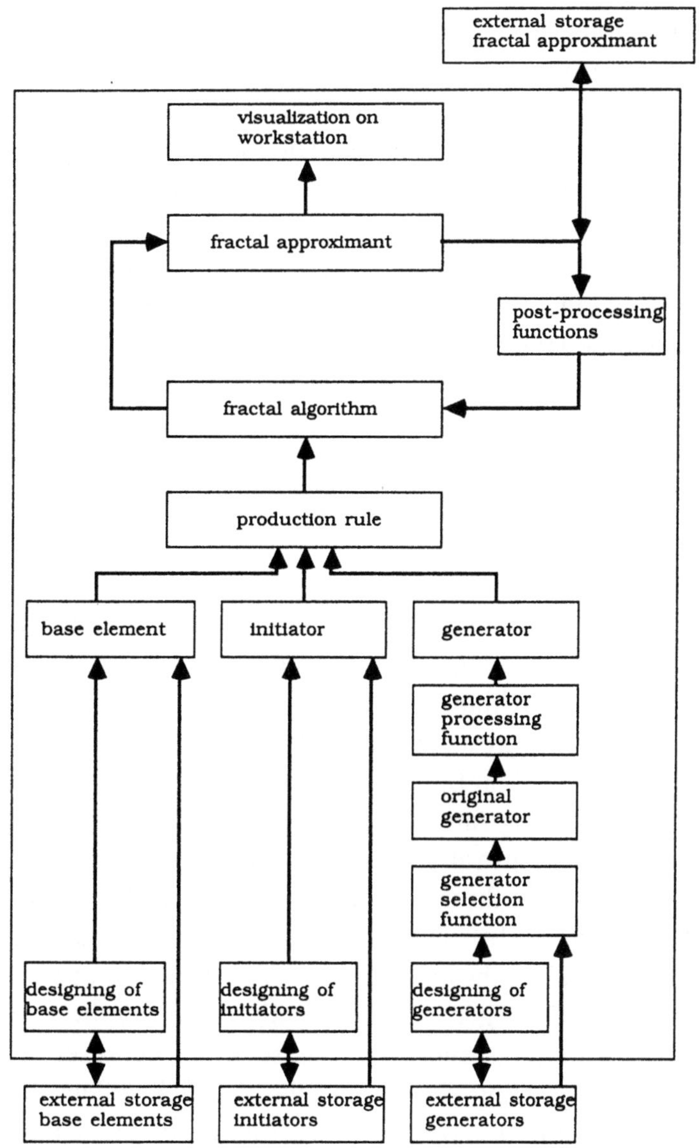

Fig. 2.32. Diagram of the 2D modelling system for iterative geometric constructions. The system itself is enclosed by a rectangle.

Interactive modelling

The modelling system is depicted in Fig. 2.32. In this diagram the system is enclosed by a rectangle. As discussed above, it is necessary to construct production rules consisting of three components: *initiator*, *generators*, and *base elements*. The three components can be designed interactively in the modelling system or taken from external storage. These external storage files contain the parameters of the rules describing the three components. Newly designed components can be added to these files.

The *generators* may contain an attribute, which refers either to a special function which processes the original *generator* or to a *generator selection function* which selects, during the calculation of the object, a *generator* from a list of *generators*. The result of one iteration step can be post-processed by a list of *post-processing functions* (these functions will be discussed in the next section). The main part of the system is the fractal algorithm, which uses the output of one iteration as input for the next one.

It is possible to define new production rules during the generation of the objects, so the ultimate object can be influenced interactively. The final result can be stored externally and this file can used to reinitialize the iteration process and to continue the calculations.

2.6.3 Modelling a Growth Process in 2D with Iterative Geometric Constructions

In this section some examples are given of simple growth processes of artificial objects. The results of the preceding section are used in models of growth processes. Although these objects are sometimes reminiscent of biological ones, the growth process has no biological significance. The examples are all based on the iterative geometric construction shown in Fig. 2.33. The resulting ramifying object is shown in Fig. 2.34. In this section it will be demonstrated how from this construction more complex objects can be built by the introduction of rules in the iteration process (see also Kaandorp 1991a). These rules differ from the rules applied in the preceding section, since they are not applied in the generation of the object but to the result of the generation procedure (an iteration step). These rules can be divided into two types: those which represent the internal properties of the growing object, and those which reflect the influence of the environment on the object. In the models of the actual biological growth processes a *generator* rule, as shown in Fig. 2.33, represents the

Internal properties of a growth process

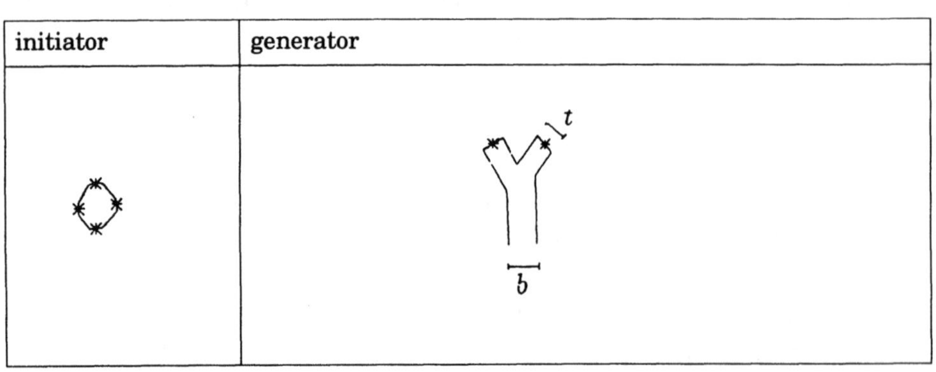

Fig. 2.33. Geometric production rule for a ramifying object. Fertile sides (sides of the preceding iteration step, which will be replaced in next iteration steps) are marked with asterisks. The resulting object is shown in Fig. 2.34. The ratio t/b (t is the size of a fertile side, b is the size of the basis of the generator) is indicated as the similarity ratio.

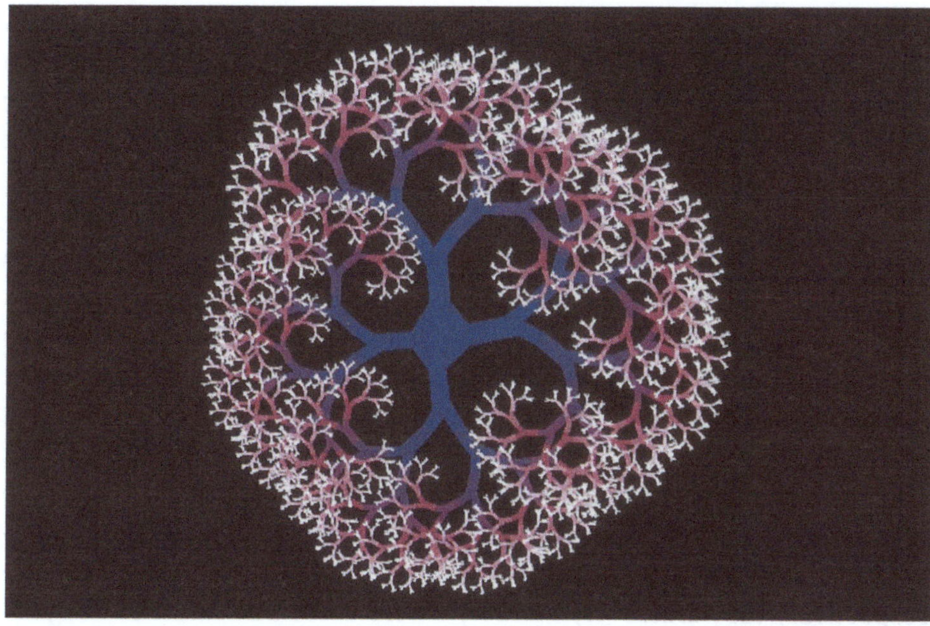

Fig. 2.34. Ramifying object resulting from the geometric production rule shown in Fig. 2.33

internal (genetically) defined part of the growth process, for example in the *generator* it is internally defined that each branch will split into two new ones. This type of rule is responsible for the self-similar aspect of the object and resides typically in the generator part of the production rule. The second type of rule represents the influence of the environment on the growth process. Examples of such rules are: the object is not allowed to grow on sites which are already occupied by the object, and the object is not allowed to grow across certain borders. In general this type of rule will disturb the regularity and the self-similar aspect of the growing object.

Influence of the environment

In the chapters on the modelling of actual biological objects, basically the same strategy will be followed as in this section on ramifying objects. First, internal rules are defined which describe the species-specific aspects, for example the architecture in which the elements of the object are connected, the properties of the transport system of nutrients through the tissue, etc. After this the influence of the physical environment is defined, for example the light distribution for models of organisms which use light as an energy source, and geometric hindrances.

Aside from being convenient demonstration objects the ramifying forms are sometimes used as a simple phenomenological description of branching patterns in biology (see Aono and Kunii 1984, Bell 1986).

The Ramifying Objects. In the production rule shown in Fig. 2.33, the ratio between the size of the base of the branch (b) and the size of the fertile tips (t) of the branches equals the self-similarity ratio. The effect of changing t/b is depicted in the series of ramifying objects in Fig. 2.35, where (t/b) is increased from 0.5 to 1.0. The effect of changing the self-similar aspect of the ramifying object causes the thin-branching and compact object A to transform gradually into a more plane-filling object C.

The first extension which can be introduced is a function which processes the original *generator* depicted in the construction in Fig. 2.33. The original generator is represented in the geometric modelling system as a sequence of transformations M_{ij}: the *generator* (see Sect. 2.6.1). The processing function is indicated as the *generator processing function*. A *generator processing function* which will generate a ramiformeous object with an irregular appearance uses the original generator as an argument and delivers a new generator on which a rotation is performed. The angle of rotation is chosen randomly between two limits (see Fig. 2.36, compare Fig. 2.24 and (2.17)). In this case for *lower_limit* and *upper_limit* respectively the values $-\pi/4$ and $\pi/4$ were used. The result of this production rule is shown in Fig. 2.37. This *generator processing function* is an example of an external influence, which disturbs the self-similar aspect of the object shown in Fig. 2.34.

In the case of the irregular ramifying object only the original generator is used as an argument in the *generator processing function*. To generate an irregular ramifying object that grows in a certain prevailing direction, some additional local information (indicated as *local_inf*) is necessary. In the present case this local information is the angle γ of a fertile element with the prevailing direction of growth. An angle θ is chosen randomly between two limits: $\gamma \leq \theta \leq \pi/2$. The probability that a specific angle is chosen is maximal for an angle which equals $\pi/2$ and minimal for an angle that equals γ, which was achieved by truncating a normal distribution with mean value $\pi/2$. An example of an irregular ramifying object growing in a prevailing direction (in this case the right upper corner) is shown in Fig. 2.38. This rule, in which branches have the highest probability to branch

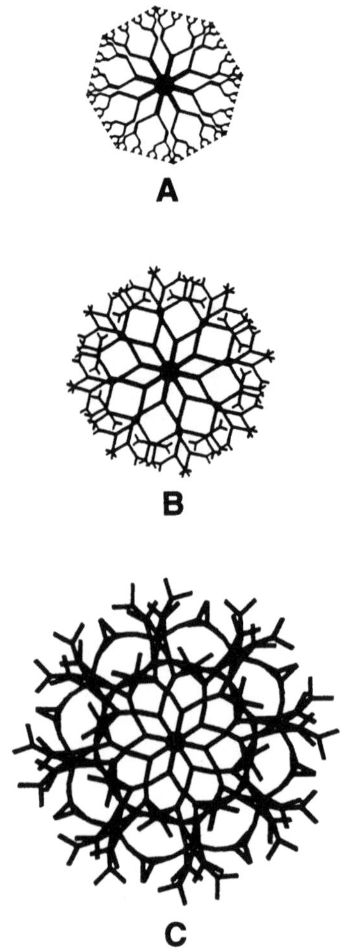

Fig. 2.35. In this series of ramifying objects the value of the similarity ratio (t/b) in the production rule in Fig. 2.33 is gradually increased in the sequence A..C. The respective values 0.5, 0.75, and 1.0 were used.

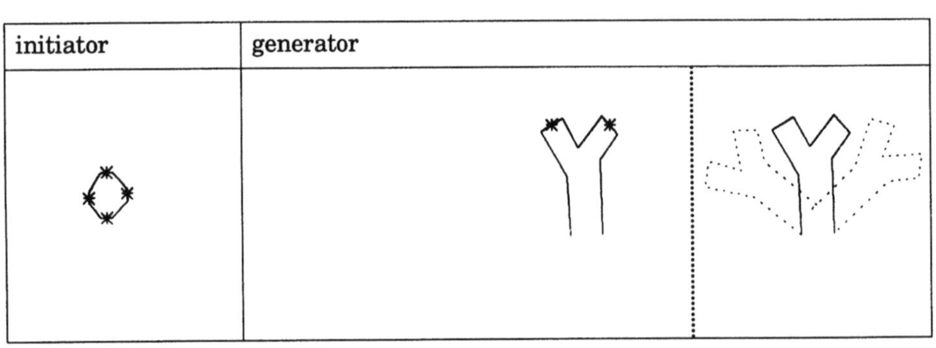

Fig. 2.36. Geometric production rule for an irregular ramifying object. The original generator is processed by a generator processing function, which allows random movements of the generator between two limits. The generator processing function is described in the right part of the generator component.

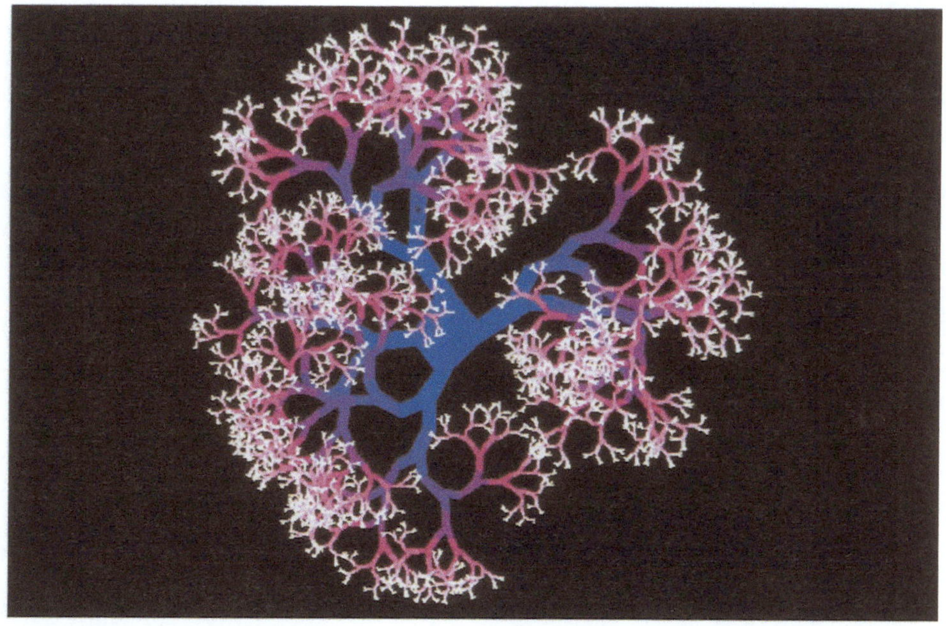

Fig. 2.37. Irregular ramifying object resulting from the geometric production rule shown in Fig. 2.36

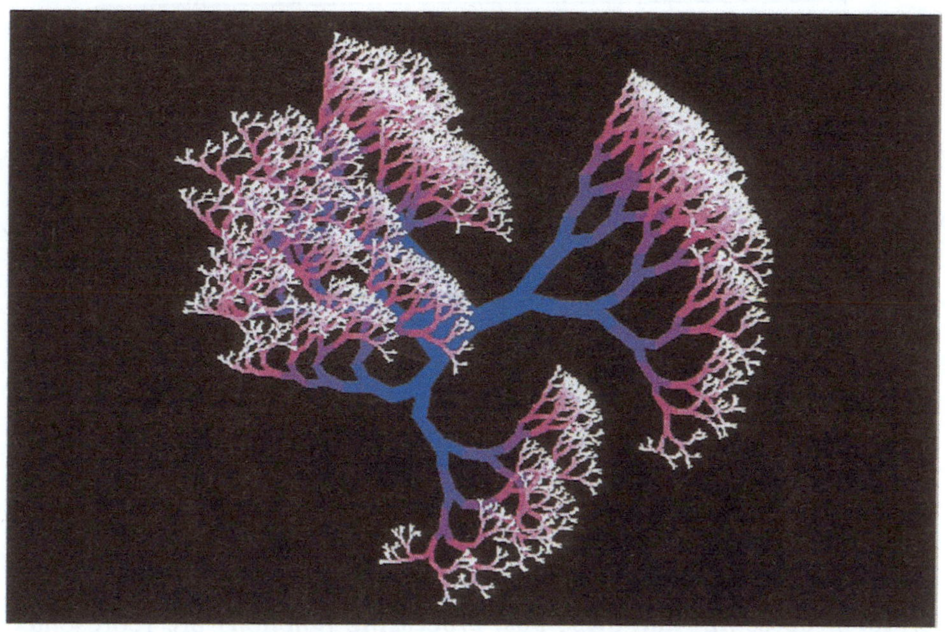

Fig. 2.38. Irregular ramifying object resulting from the geometric production rule shown in Fig. 2.36, growing towards a prevailing direction

towards the right upper corner, is an example of an external influence on the growth process acting non-uniformly on the growing object. A biological example is a light source influencing the growth process. Local information is necessary to identify the position of a growing element with respect to the light source.

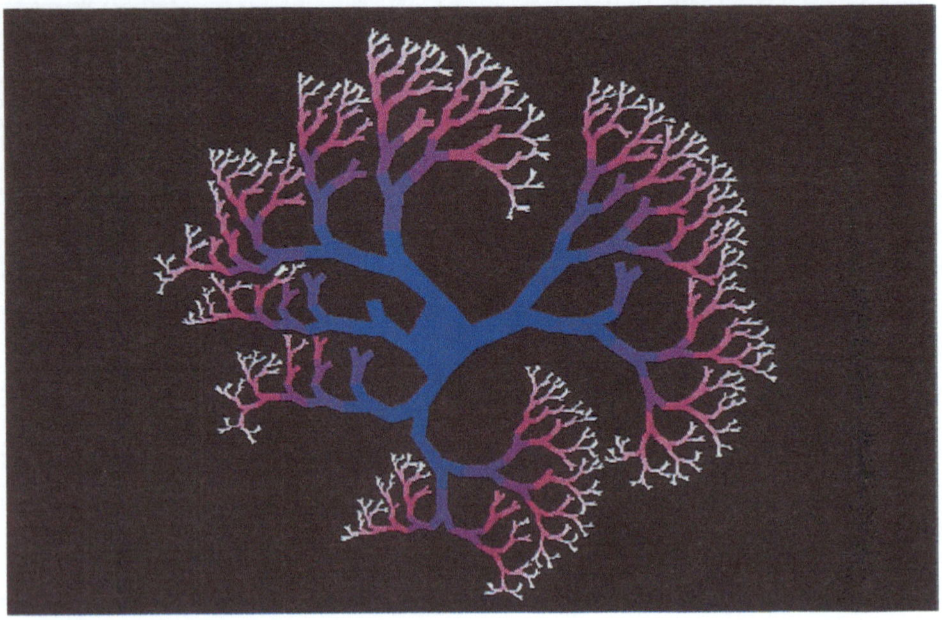

Fig. 2.39. Irregular ramifying object in which intersecting branches of the object shown in Fig. 2.38 are removed by a *post-processing function*

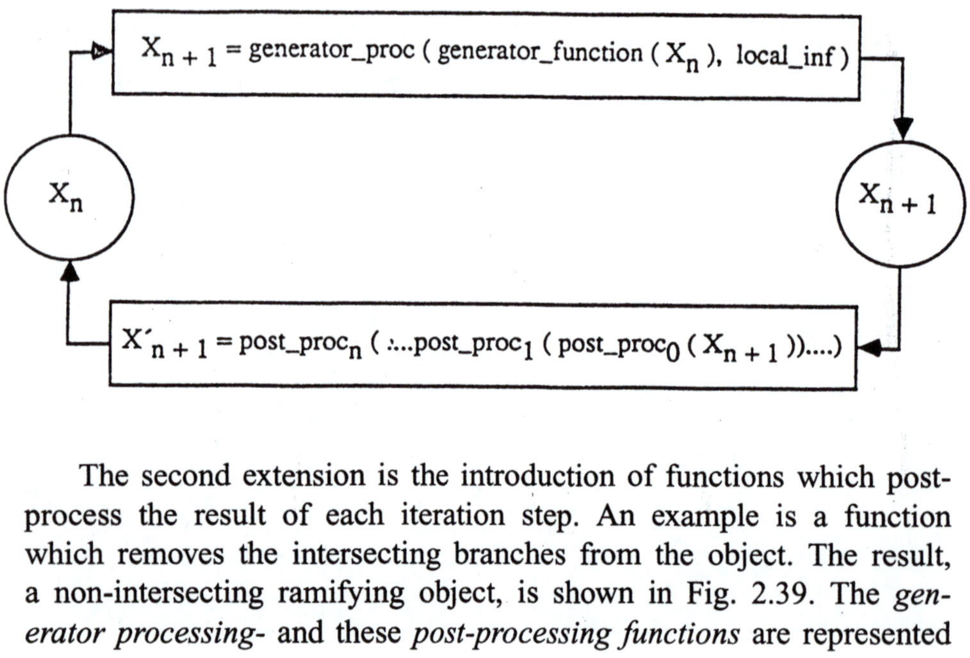

Fig. 2.40. Diagram of the iteration process in which three extensions are introduced in the original iteration process (Fig. 2.2): the *generator processing function* (*generator_proc*), a chain of *post-processing functions* (*post_proc$_x$*), and local information (*local_inf*)

Post-processing functions

The second extension is the introduction of functions which post-process the result of each iteration step. An example is a function which removes the intersecting branches from the object. The result, a non-intersecting ramifying object, is shown in Fig. 2.39. The *generator processing-* and these *post-processing functions* are represented in a new diagram of the iteration process: Fig. 2.40 shows a chain of *post-processing functions* (*post_proc$_0$*, *post_proc$_1$*, .., *post_proc$_n$*).

Non-intersection rule

The non-intersection rule is an externally defined geometrical restriction which influences the growth of all biological organisms. This rule will later become one of the major rules in the growth models of actual biological organisms. Non-intersection can be introduced as an additional condition in the edge replacement system describing the object from Fig. 2.39:

$$
\begin{aligned}
\mathit{base\ element} &= edge(V_a, V_b) \qquad (2.26)\\
\mathit{initiator} &= ((edge(V_0, V_1), F); (edge(V_1, V_2), NF);\\
&\quad (edge(V_2, V_3), F); (edge(V_3, V_4), NF);\\
&\quad (edge(V_4, V_5), F); (edge(V_5, V_6), NF);\\
&\quad (edge(V_6, V_7), F); (edge(V_7, V_0), NF);), age = 0);\\
\mathit{generator\ processing}&\\
\mathit{function}(\gamma) &= \text{for each } (edge(V_i, V_{i+1}), F) \text{ a rotation}\\
&\quad \text{matrix } R_\theta \text{ is determined for}\\
&\quad \gamma \leq \theta \leq \pi/2; \theta \text{ is chosen randomly}\\
&\quad \text{between the two limits}\\
&\quad P(\theta) \text{ is a (truncated) normal}\\
&\quad \text{distribution with } \overline{\theta} = \pi/2\\
\mathit{new\_edges} &= \{(edge(V_i, R_\theta(M_{1j}(V_i)), NF);\\
&\quad (edge(R_\theta(M_{1j}(V_i)), R_\theta(M_{2j}(V_i))), NF);\\
&\quad (edge(R_\theta(M_{2j}(V_i)), R_\theta(M_{3j}(V_i))), F);\\
&\quad (edge(R_\theta(M_{3j}(V_i)), R_\theta(M_{4j}(V_i))), NF);\\
&\quad (edge(R_\theta(M_{4j}(V_i)), R_\theta(M_{5j}(V_i))), NF);\\
&\quad (edge(R_\theta(M_{5j}(V_i)), R_\theta(M_{6j}(V_i))), F);\\
&\quad (edge(R_\theta(M_{6j}(V_i)), R_\theta(M_{7j}(V_i))), NF);\\
&\quad (edge(R_\theta(M_{7j}(V_i)), V_{i+1}), NF); \}\\
\mathit{generator} &= \text{if (an edge in } \mathit{new\_edges} \text{ intersects}\\
&\quad \text{an edge of the object) then}\\
&\quad\ (edge(V_i, V_{i+1}), F); \rightarrow (edge(V_i, V_{i+1}), NF);\\
&\quad \text{else}\\
&\quad\ (edge(V_i, V_{i+1}), F); \rightarrow \mathit{new\_edges}\\
&\quad (edge(V_i, V_{i+1}), NF); \rightarrow (edge(V_i, V_{i+1}), NF);
\end{aligned}
$$

In this replacement system the angle γ is an argument of the *generator processing function* and the newly generated edges are only added to the ramifying object in the case that this does not lead to intersection.

Geometric restrictions

The chain of *post-processing functions* may contain all kinds of rules for the growing object. Without these *post-processing functions* the elements of a ramifying object cannot interact with each other; they make it possible to introduce restrictions, for example that an element is not allowed to intersect an other element, or elements should remain at a certain distance from each other.

In Fig. 2.39 growth stops as soon as a branch is going to intersect a part of the object. In Fig. 2.41 the iteration process of Fig. 2.40 is extended with an additional cycle. In this new iteration process a fertile side produces a new branch that is tested for intersections with the object. When an intersection is found, further new branches are generated until a branch is found, by trial and error, that does not intersect, or the number of attempts exceeds a certain maximum (*nretry*). In the latter case the branch is removed from the object and the tested fertile side becomes non-fertile. In Fig. 2.42 the result of this new addition is shown. In this figure each fertile side produces a maximum number of new branches (in this picture *nretry* = *10* was used); the result is that branches try to avoid each other and the number of branches created is higher than in Fig. 2.39. For the generation of Fig. 2.43 a new post-processing rule is inserted in the chain of post-processing rules, stating that the ramifying object is not allowed to intersect a nearly closed box. The aperture of this box is found by the ramifying object by trial and error.

Avoiding intersections by trial and error

In actual biological objects the intersections as occurring in Fig. 2.37 will appear less frequently. The objects can have a self-avoiding architecture. It is possible to construct a branching generator which exactly avoids itself in each iteration step as shown in Fig. 2.44 (some more examples of self-avoiding ramifying structures are given by Mandelbrot 1983, Aono and Kunii 1984). Of course this mathematical property of self-avoidance and strict self-similarity (not considering the trunk of this object) would quickly get lost in an actual growth process, where the physical environment will influence the growth process (examples of this will be shown in the next chapter). Only statistical self-similarity remains in biological ramifying structures. Growing branches in seed plants suppress growth

Self-avoidance

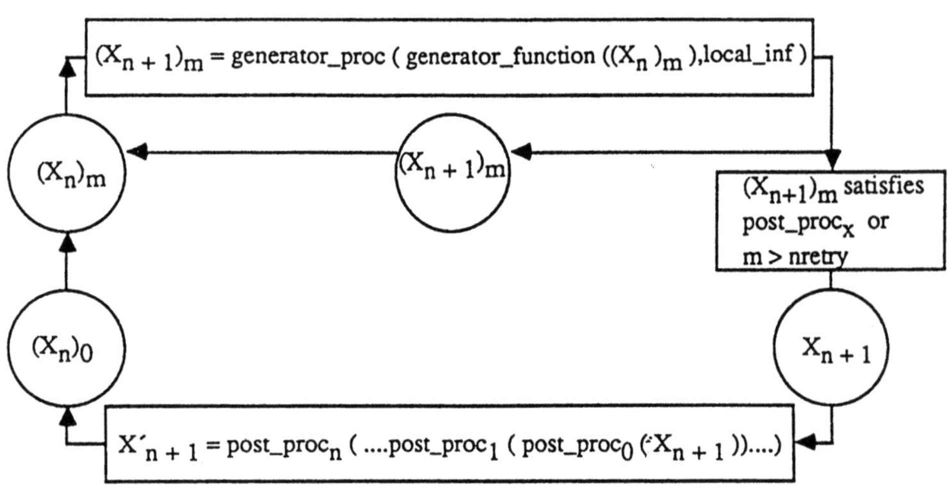

Fig. 2.41. Diagram of the iteration process, where the iteration process (Fig. 2.40) is extended with an additional cycle

Fig. 2.42. Irregular ramifying object where all fertile sides generate a maximum of *nretry* branches until a branch is found that does not intersect the object (see Fig 2.41)

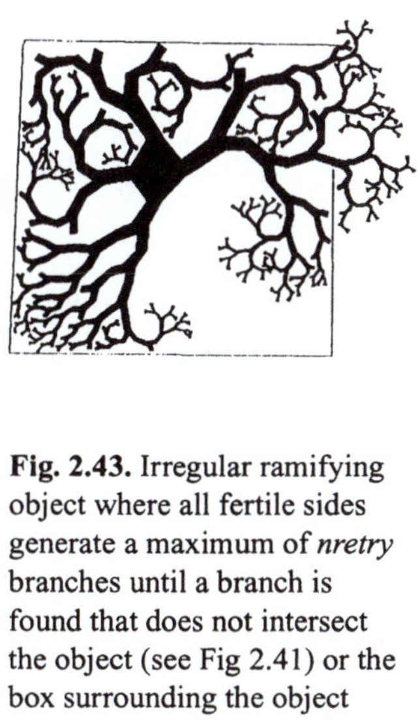

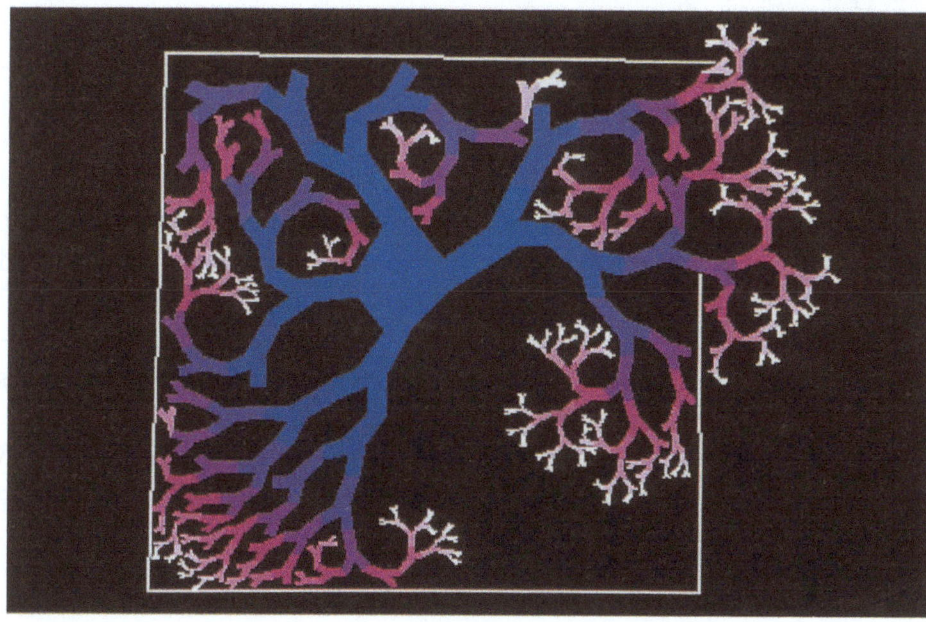

Fig. 2.43. Irregular ramifying object where all fertile sides generate a maximum of *nretry* branches until a branch is found that does not intersect the object (see Fig 2.41) or the box surrounding the object

in other branches by apical dominance where the phytohormone auxin, produced by the top meristem, suppresses growth in the side branches. In many autotrophic[1] branching organisms (e.g. seed plants, many stony

[1] A (photo-)autotrophic organism requires only inorganic substances and light as an energy source for growth.

Fig. 2.44. Self-avoiding ramifying object in which a branching generator is chosen which exactly avoids the other branches in new iteration steps

corals) growth in the lower branches is suppressed by the cast shadows of the upper branches. In many branching heterotrophic[2] organisms (e.g. some bacteria colonies, bryozoans, many sponges), as well in many branching growth patterns in physics, growth in the other branches is suppressed simply by depletion of the nutrients necessary for growth (see for example Fig. 2.15). In DLA-like structures, as electrochemical deposition, dielectric breakdown patterns, etc. (see Sect. 2.4) this suppression is very effective: loops are very seldom observed. The first three reasons (self-avoidance, apical dominance and cast shadows) can explain why the phenomenon of anastomosis is so surprisingly rare among most seed plants. Probably the growth process is highly controlled among seed plants and anastomosis only occurs when the process is disturbed strongly. Among the root systems of the seed plants this phenomenon is observed much more frequently; one possible explanation could be that the physical environment disturbs the growth more in combination with a more complicated systems of meristems (in the meristem in the root, which causes secondary growth, the cambium is star-shaped instead of ring-shaped). This system seems to be apt to generate highly irregular forms. In Chap. 3 it will be demonstrated that anastomosis appears more frequently among marine sessile organisms.

Anastomosis

[2] A heterotrophic organism requires an external source of one or more organic compounds as an energy source. In many marine organisms these compounds are often obtained by filtering suspended material from the environment.

2.7 A Review of the Methods

The main objective of this book is to gain insight into the morphogenesis of organisms and the influence of the physical environment on the emergence of form. For this purpose reaction diffusion mechanisms are less applicable. The reaction diffusion approach is a model for the prepattern, consisting of a spatial pattern of activator and inhibitor concentrations. In a model of the emergence of forms which may result from this prepattern an additional morphological model is necessary. It can be expected that it is a fruitful approach to introduce this method in morphological models.

Iterative geometric constructions will be the prevailing method for constructing models of growth processes in biology. This is conceptually a very natural way to describe a growth process. Basically it is always possible to find an iterative geometric construction which mimics a certain growth process, for example the process shown in Fig. 2.6. In this case a set of base elements (the base element is the geometrical description of a polyp) replaces a base element in the iteration process. In the actual growth process these polyps will exhibit many types of interactions.

The actual growth process is a non-deterministic process, partly because it is influenced continuously by the environment (for example the supply of nutrients) and because growth is limited by geometric restrictions. A very obvious geometric restriction is that growth cannot take place at a site in space which is already occupied by the growing object. This type of limitation and the influence of the environment cannot be described with a L-system.

In the discussion of iterative geometric constructions it was demonstrated that it is possible to introduce geometric restrictions in the model (see the edge replacement system in (2.26)). It is also possible to include models of the physical environment in the iterative geometric constructions, more detailed examples of which will be given in the next chapter.

The DLA model is a successful model in mimicking the growth process of an aggregation of loose particles and the influence of the environment (the nutrient source). As a model of a growing coherent structure, DLA is less applicable. In Chap. 3, however, it will be shown that it is possible to combine both approaches (the iterative geometric constructions and the DLA model) in order to take advantage of both.

The IFS method is basically unsuitable, since it only delivers an image of the object and not a complete simulation of its growth process. The method is a suitable approach for describing the self-similar aspects of the image.

3 2D Models of Growth Forms

This chapter discusses how a system of rules can be created, as shown in the section on the ramifying objects, suitable for the simulation of the growth process of various sessile marine organisms, for example sponges (Porifera) and corals (Scleractinia). A crucial difference with the example of the ramifying objects is that each modelling step is supposed to have a biological significance. An important reason to use sponges and corals as subjects of a case-study is that these organisms exhibit a relatively simple growth process, which makes it comparatively easy to design geometric production rules to simulate growth processes. Although the marine sessile organisms belong to many very different taxonomic groups, it is possible to distinguish several types of corresponding growth processes within these groups, in which a similar architecture emerges. First almost all groups belong to the large group of modular organisms. In this group there is a subset of organisms which are formed by one type of growth process, which will be discussed in particular in the next sections. This growth process can be found within sponges, stony corals and many other marine sessile organisms and will be indicated as radiate accretive growth. Modular growth and radiate accretive growth are the first subjects in this chapter, followed by a 2D model of radiate accretive growth.

3.1 Modular Growth

Modular growth is defined in Harper et al. (1986) as the growth of genetic individuals by repeated iteration of (multi-cellular) parts, the modules. Modules might be the polyp of a (stony) coral or an octocorallian, a zooid in a bryozoan, a hyphe in a fungal colony, a shoot with an apical meristem in seed plants, or an osculum together with its aquiferous canals in sponges. A characteristic of modular organisms is that they do not have a determinate form, although the module itself might have a determinate form (in sponges the module is variable in form). Modular organisms contrast with unitary organisms, in which a single-celled stage, usually a zygote, develops into a determinate structure. As well as a great plasticity in form, modular organisms share such properties as the possibility to survive partial mortality and the absence of a physiologically limited age. Modular growth and architecture is a very general principle in biology (see also Larwood and Rosen 1979, Jackson et al. 1985, Vuorisala and Tuomi 1986, Rubin 1987), which is also applicable in other sciences, e.g. computer science (see Unger and Bidulock 1981, on the modular design of multi-computer systems).

Unitary organisms

The modular growth of many sponges and corals is relatively simple when compared to more complex modular organisms like seed plants. For many sponges it can be defined as parallel modular growth (Kaandorp 1991b), where the various modules grow almost independently, only limited by steric hindrance. Because of the almost independently growing modules, which are not limited by the development of other modules, some important simplifications can be made in the modelling of the growth process. For organisms with non-parallel growth, such as seed plants, the growth of the modules, the apical meristems, is limited by the development of other modules, for example the root system. A consequence of the parallel modular growth is that the organisms can increase in size theoretical without limit and without a decrease in growth velocities. In reality, growth of these organisms will be limited by external factors like strong water movements.

Parallel modular growth

In parallel modular growth the final shape is developed by almost independently growing modules. It is the integrated result of the growth process of the individual modules and the influence of the environment. There is no overall control in the final shape of the organism, as in seed plants. In the latter this overall control becomes manifest for example in the phenomenon of apical dominance, where the top meristem suppresses growth in the lower meristems by secreting the phytohormone auxine. As a consequence of a lack of overall control the final shape often shows a clear response to the environment. Organisms with parallel modular growth exhibit in general a larger plasticity in forms, higher degrees of irregularity, and anastomosis compared to non-parallel modular organisms.

Apical dominance

3.2 Radiate Accretive Growth

Radiate accretive growth can be defined as an iterative growth process in which layers of material are added externally to the tip of a preceding growth step, which remains unchanged in the next growth steps. In this

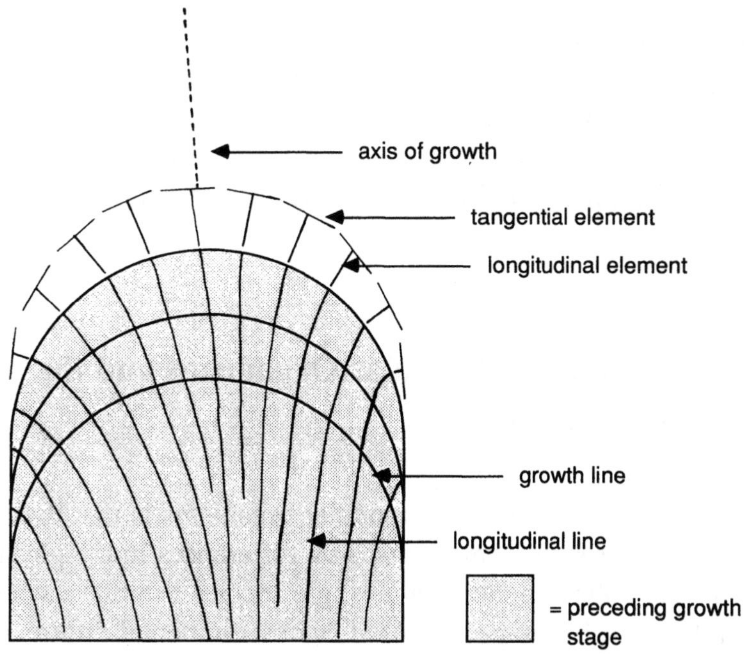

Fig. 3.1. A diagram of a radiate accretive growth process: new layers consisting of tangential and longitudinal elements are deposited upon the preceding growth stage. The tangential elements correspond to the growth lines and indicate the circumference of the organism in earlier growth stages.

process the thickness of the layers is highest at a minimal angle between a tangential element and an axis of growth (see Fig. 3.1) and decreases towards the sides of the tip. In this process a typical radiate architecture is formed, where the longitudinal elements are set perpendicular on the preceding tangential elements. This type of growth can be found among members of various groups of modular marine organisms: stony corals (Graus and Macintyre 1982), sponges (Wiedenmayer 1977, Kaandorp 1991b), coralline algae (Bosence 1976), and a symbiotic octocoral sponge association (Van Soest and Verseveldt 1987). A distinction can be made between the autotrophic organisms where the growth process is mainly light-dependent and the heterotrophic ones where growth is mainly determined by the supply of nutrients, suspended material in the water, from the environment. Combinations of the autotrophic and heterotrophic source of carbon for the metabolic synthesis are also possible. In many scleractinians, photosynthesis of the zooxanthellae is the most important part of the metabolism, while zooplankton feeding represents only an insignificant component of the energy intake (Porter 1974, Spencer Davies 1984, Edmunds and Spencer Davies 1989). For some species it is even demonstrated that no zooplankton at all is needed for tissue growth (Franzisket 1969, Johannes 1974). Zooplankton may, however, provide essential elements such as N and P for the production and maintenance of tissue (Bythell 1988). Although the radiate accretive growth process is a relatively simple one it may result in a large variety of growth forms, as shown

Autotrophic and heterotrophic organisms

in the following sections. In a general model of radiate accretive growth the influence of the environment on the growth process will consist of two components: the influence of the light intensity and the supply of nutrients. The last component is closely related to the exposure to water movement, a dominant environmental parameter for most of the marine sessile organisms.

3.3 Growth Forms of Modular Organisms and the Physical Environment

A typical characteristic of many modular organisms is that they exhibit a wide range of growth forms, caused by differences in the physical environment. This phenomenon that a single species exhibits a wide range of growth forms (phenotypes), caused by environmental differences, is a well-known problem in the taxonomy of sponges, corals, and other modular organisms. It is often possible to arrange the growth forms along a physical gradient, so that the forms gradually transform into each other with the changing environmental parameter.

Fig. 3.2. The sponge *Haliclona oculata* under natural conditions (the photograph was made in Eastern Scheldt in the Netherlands by M.J. de Kluijver).

An example of a heterotrophic organism with radiate accretive growth is the sponge *Haliclona oculata* (see Hartman 1958, De Weerdt 1986). In Fig. 3.2 a picture of this sponge under natural conditions is shown. The overall body shape of this sponge tends to be more or less flattened, although forms which are branching in all directions often appear as well. In Fig 3.3 examples from the range of growth forms are shown. Basically these vary within a range from quite regular thin-branching (form A) to irregular plate-like growth forms (form D). The thin-branching form is typical for sheltered conditions (the displayed sample was collected in a tide-less salt water lake), while the plate-like form is found under conditions with more exposure to water movement. In general the thickness of the branches and the branching rate increase with the rate of exposure to water movement. Forms A and D are two extremes, between which all kind of intermediates, such as B and C, can be found. Deviations from this general trend easily arise as a result of damage to the sponge in the course of time. More irregular forms are found among sponges which have an age of several years (see Fig. 3.3E). Probably tissue-material is removed by abrasion or partial mortality and irregularities arise when new material is added. Another tendency which can be observed in this range is an increase in irregularity. These aspects of the growth forms will be discussed in more detail in the section on the comparison of forms. This

Fig. 3.3. Range of growth forms of the sponge species *Haliclona oculata*. Form A is typical for sheltered sites, in the range A - D the exposure to water movement increases, and form E is an irregular form with the age of several years.

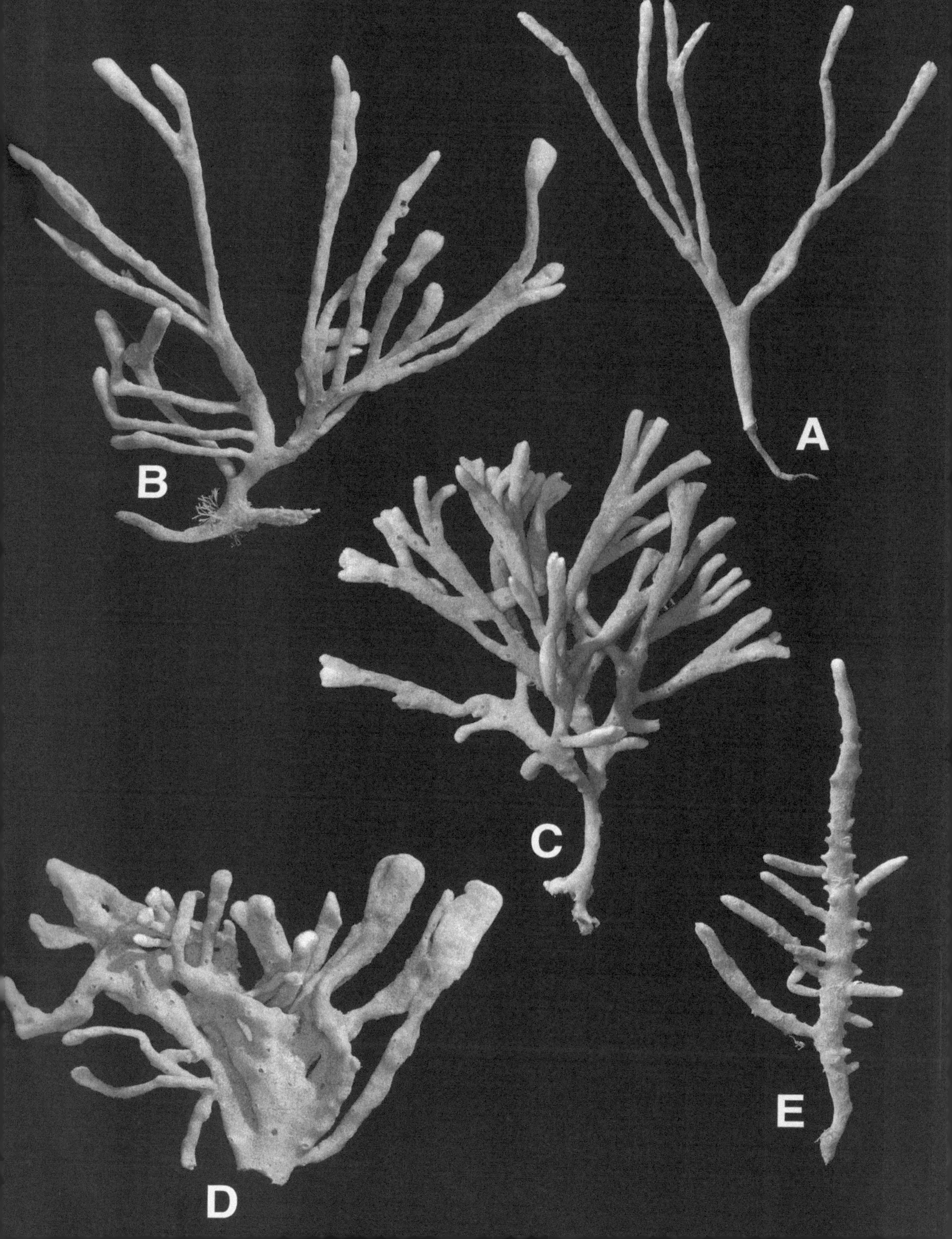
A
B
C
D
E

Fig. 3.4. The stony-coral *Montastrea annularis* (photograph made by W.H. de Weerdt in the Caribbean area)

large plasticity in growth forms, caused by differences in exposure to water movement, is a well-known phenomenon in sponges (see e.g., Bidder 1923, Warburton 1960, Barthel 1991).

The growth forms of *H. oculata* are something between the "plates" and "trees" of the basic forms (for marine sessile animals) described by Jackson (1979). Such a division in basic growth forms is very artificial, as can be seen in the example of *Haliclona oculata* and in more examples which will be shown in this section. In reality these species can be arranged in a series of ecotypes along a gradient of one or more environmental key parameters.

An example of a (mainly) autotrophic organism with radiate accretive growth is the scleractinian coral *Montastrea annularis*. In Fig. 3.4 a photograph of this coral under natural conditions is shown. For a mainly light-dependent organism, the range of growth forms varies with the light intensity. This species exhibits a hemispherical colony form under circumstances with a maximum light intensity, when the colony grows close to the water surface. The colony gradually transforms from hemispherical through column-shaped and tapered forms to a substrate covering plate (see Fig. 3.5) when the light intensity decreases (see Graus and Macintyre 1982; see also Roos 1967 about the stony coral *Porites astreoides*).

Organisms for which light as well as the heterotrophic nutrient source are the main environmental parameters determining the growth form include many of the Scleractinia, for example *Acropora palmata* (see Bythell

→

Fig. 3.5. Range of colony shapes of stony coral *Montastrea annularis*. The colony gradually transforms from hemispherical (A), column-shaped (B), and tapered forms (C) to a substrate covering plate (D) when the light intensity decreases.

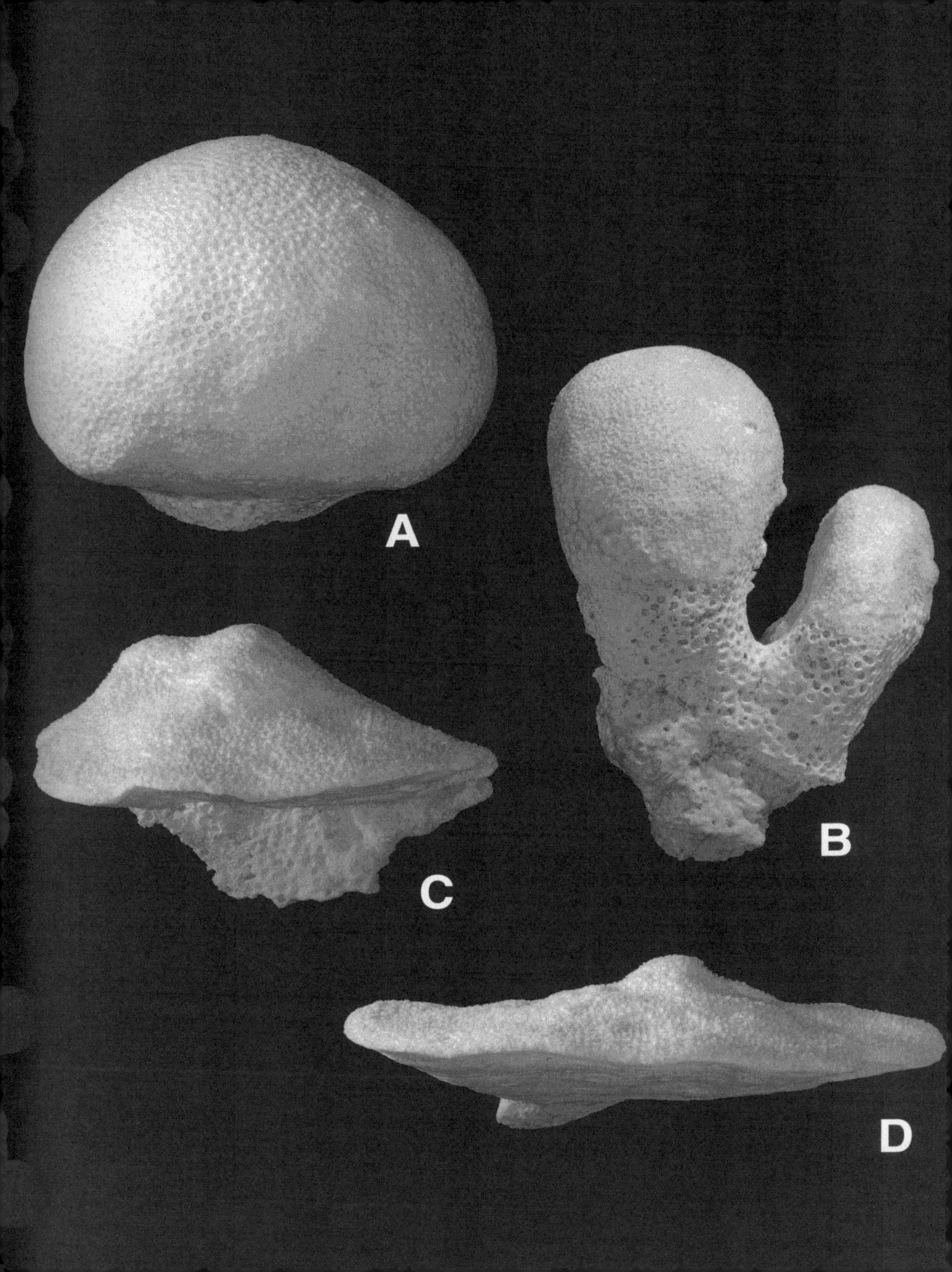
A
B
C
D

Fig. 3.6. Growth form of the stony-coral *Acropora palmata* (the photograph was made by W.H. de Weerdt in the Caribbean area)

Fig. 3.7. Growth forms of the hydrocoral *Millepora alcicornis* (photograph made by W.H. de Weerdt in the Caribbean area)

1988), and some Porifera (see Wilkinson et al. 1988). In Fig. 3.6 some growth forms of this stony-coral are shown. Theoretically one would expect, instead of an one-dimensional variation in forms as in the two preceding examples, a two-dimensional variation in forms. In most cases one of the environmental parameters will be the dominant cause of the variation in forms. In Veron and Pichon (1976) beautiful ranges in growth forms of the scleractinian corals *Pocillopora damicornis* and *Seriatopora hystrix* are shown, where in both cases the growth forms transform from compact branching forms to thin-branching in the gradient from exposed to water movement to sheltered conditions. An example of the range of growth forms of the hydrocoral *Millepora alcicornis* (see De Weerdt 1981), using the autotrophic as well as the heterotrophic energy source, is shown in Figs. 3.7 and 3.8. This species shows, in a range from shallow to deeper

Fig. 3.8. Range of colony shapes of the hydrocoral *Millepora alcicornis*. The colony gradually transforms from plate-like growth (A) to thin branching forms (D), when exposure to water movement decreases.

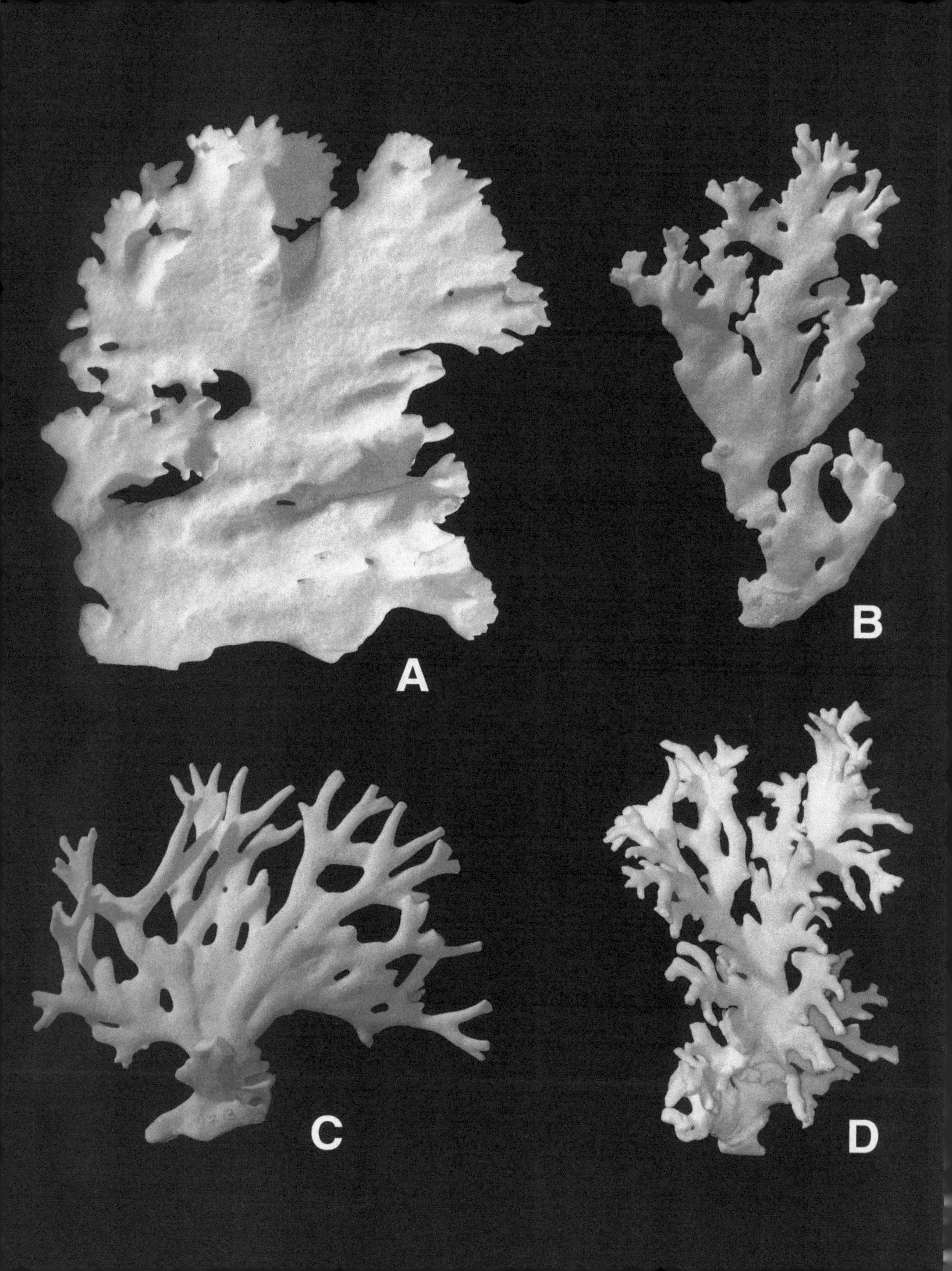
A
B
C
D

growth sites, a gradual transition from plate-like growth forms to thin-branching forms. In the same range, both key parameters, exposure to water movement and light, are varying, water movement being the main parameter influencing the growth form. The growth process of *Millepora alcicornis* is slightly different from the radiate accretive growth process, and is comparable with *Millepora complanata* (see Lewis 1991).

For reasons of simplicity, the discussion of the biological significance of the rules in the development of the models will be limited to the two examples of organisms with radiate accretive growth given above: the autotrophic stony coral *Montastrea annularis* and the heterotrophic sponge *Haliclona oculata*.

3.4 Description of the Internal Architecture of the Autotrophic Example: *Montastrea annularis*

The skeletal architecture of many corals can be visualized by sectioning the colony. If a slab is taken from such a section and x-rayed, and a positive print is made of the negative, it is possible to trace the growth process morphologically. In Fig. 3.9 a longitudinal section is made through the crest of the colony along the axis of growth (which for this species is vertical). The annual growth is visible as dark and light density bands in x-radiographs (see Graus and Macintyre 1982) and it is possible to distinguish growth lines, which correspond to the tangential lines in Fig. 3.1.

Annual growth of Montastrea annularis

The tangential and longitudinal lines in Fig. 3.9 correspond to the faces of the corallites, the cups containing the polyps of the colony. A clear radiate accretive structure can be seen: growth is the strongest at the crest of the colony and decreases towards the sides. The longitudinal faces of the corallites are set perpendicular to the preceding layers. The preceding growth stage remains unchanged in the growth process. This is very obvious for stony corals where only the surface is alive and where the living polyps are depositing material upon the dead core. The growth process of *Montastrea annularis* can also be followed experimentally by staining living colonies with Alizarine Red S (see Graus and Macintyre 1982). The surface of the colony at the moment of staining can be reconstructed from the coloured band which can be seen in sections used for the x-radiographs. A tangential view of the colony shows an arrangement of the corallites, which can be described, in a simplified version, as a regular tessellation of equal-sided pentagons and hexagons (see Fig. 3.10A and Fig. 3.10B). The tangential edges of the pentagons and hexagons have

Tangential arrangement of the corallites

Fig. 3.9. A longitudinal section through a column-shaped colony of the stony coral *Montastrea annularis* with a radiate accretive growth process. The insertion of a tangential element and the emergence of a new longitudinal line is marked with an arrow.

A

B

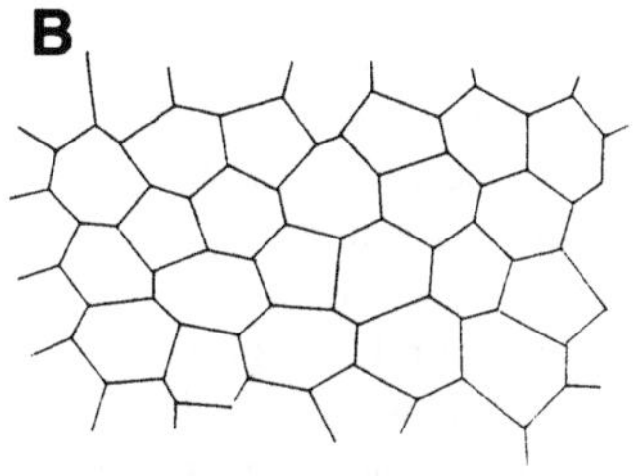

Fig. 3.10. (A) Tangential view of a colony of the stony coral *Montastrea annularis*
(B) Diagram of the tangential view of a colony of the stony coral *Montastrea annularis* displayed in (A) showing a predominantly pentagonal and hexagonal arrangement of the corallites

about the same size, while in x-radiographs a clear variation in the length of the longitudinal edges can be seen (see Fig. 3.9) over the surface of the colony.

The polyps are interconnected and can locally transport organic compounds and calcium to the neighbouring polyps (Taylor 1977, Rinkevich and Loya 1983). This local transport system can sustain the energy intake of an individual polyp, but is only developed in a limited way. At the borders of the colony, where the longitudinal edges make a maximal angle with the vertical, the energy intake by photosynthesis of the polyp together with the support of the neighbouring polyps exceeds a critical. value after which growth is not possible anymore. The clusters of modules (the polyps) of *Montastrea annularis* grow independently from each other, there is no overall control of the growth process, and the growth of the clusters of modules can be described as a parallel process.

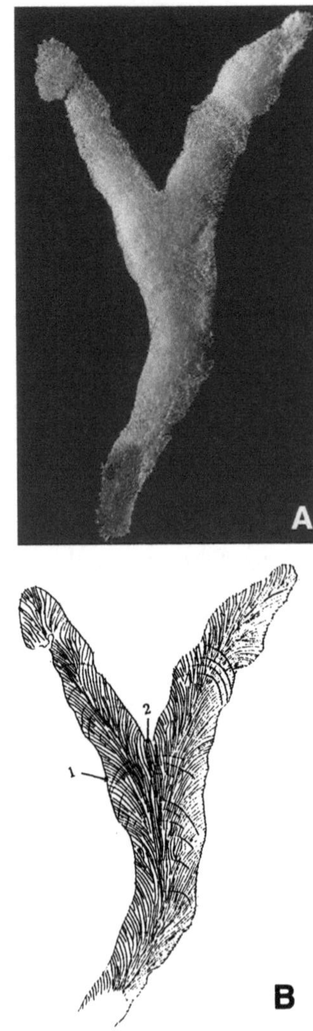

Fig. 3.11. (A) A longitudinal section of a branching sponge *Haliclona oculata* with a radiate accretive growth process as seen through a stereo microscope. (B) Drawing of the longitudinal section shown in (A). The longitidinal and some of the tangential fibres are shown in a line drawing. The insertion of a tangential element and the emergence of a new longitudinal line is marked (1). The deletion of a tangential element, causing the disappearance of longitudinal line in the growth process, is marked (2).

3.5 Description of the Internal Architecture of the Heterotrophic Example: *Haliclona oculata*

Among the members of the class of Demospongiae only species with a certain kind of skeleton architecture can develop erect tree-like growth forms. The skeleton of *Haliclona oculata* consists of discrete identical skeleton elements (the spicula) which are connected by spongin and consolidated in a 3D mesh, where a distinction between ascending and interconnecting fibres of spicula can be made. This type of skeleton is known as "regular anisotropic reticulate" and the type of architecture as "radiate accretive" (terminology after Wiedenmayer 1977). Demospongiae with a halichondrid skeleton (cf. Wiedenmayer 1977), where the spicula are oriented randomly as found for example in *Halichondria panicea*, usually develop quite irregular (often encrusting) growth forms and seldom exhibit tree-like forms. The skeleton can be made visible by drying and sectioning the sponge. The effect of drying is that all material, except the spicula, virtually disappears by shrinking. An overall view of the skeleton, as seen through a stereo microscope, is shown in Fig. 3.11A. In the anisotropic skeleton ascending (longitudinal) fibres and interconnecting (tangential) fibres can be discriminated. The radiate accretive architecture is the reflection of a growth process in which a new layer of material is added at the tip of a branch or column. The tangential fibres correspond to surfaces of earlier growth stages. In Fig. 3.11B the longitudinal and some of the tangential fibres are shown in a line drawing of Fig. 3.11A.

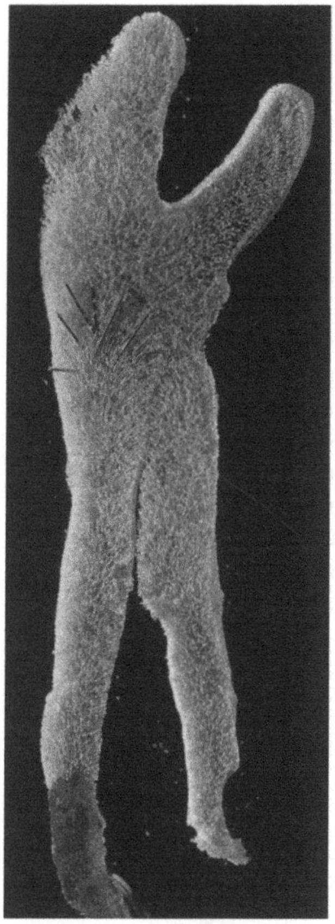

Fig. 3.12. A longitudinal section of a tip of *Haliclona oculata*, in which the surface of a preceding growth stage is marked with needles. The growth lines can be reconstructed by interconnecting the ends of the needles (1.5 month experiment).

The growth process of *Haliclona oculata* can be followed experimentally by marking experiments. The surface of the sponge can be marked with minute stainless steel needles. The needles are pushed into the living sponge, the ends of the needles corresponding with the original surface. The growth lines can be reconstructed by interpolating the ends of the needles (Kaandorp and De Kluijver 1992). In longitudinal sections through the marked tips the growth process can be retraced. An example of such a section is shown in Fig. 3.12.

A microscopic view of the skeleton is presented in Fig. 3.13. From the tangential view, Fig. 3.13A, the structure of the tangential fibres is apparent. The spicula are arranged in 4- to 6- (seldom 3-) sided polygons. The length of a side of a polygon is about the size of one spiculum. The surface of the sponge is tessellated with such polygons. In Fig. 3.13B an idealized version of the skeleton is presented. In this figure the longitudinal bundles situated in the same plane are shown, with three layers of the tangential polygons partly visualized in order to demonstrate the coherence of the longitudinal bundles and tangential layers. From the longitudinal view it can be seen that the spicula of the longitudinal fibres are arranged in bundles which are about two spicula thick and form fibres of variable length. The 3D mesh of spicula in a tip of this sponge possesses a radial symmetry: a longitudinal section (parallel to the axis of the tip) will always show about the same structure; the tips however may only be somewhat flattened (Fig. 3.3C and D).

The aquiferous system of *Haliclona oculata* is poorly developed, when compared to a related species like *Haliclona simulans* (see Johnston

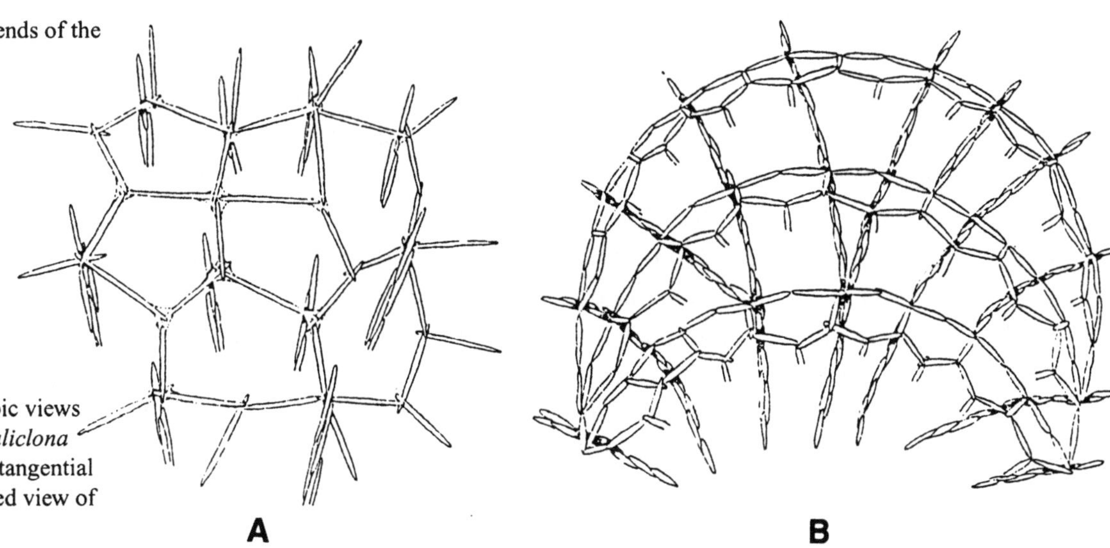

Fig. 3.13. Microscopic views of the skeleton of *Haliclona oculata*: view A is a tangential section, B an idealized view of a longitudinal one

1862, De Weerdt 1986). Only close to the oscula (the exhalant apertures of the sponge) is macroscopic evidence of canals found (see Fig. 3.14).

In *Haliclona simulans*, which also exhibits erect growth forms with radiate accretive growth, the aquiferous system is far more evolved and visible as an extensive system of canals. Probably the aquiferous system of *Haliclona oculata* is strongly supported by external water movements as well (compare Vogel 1974). Under conditions with strong water movements plate-like growth forms are possible, whereas under sheltered conditions a decrease of food supply will occur in the tissue, unless it is in short distance contact with the environment; in this case only thin-branching forms will occur. In the related species *H. simulans*, with a more evolved aquiferous system, relatively wide branches and more globular forms are found (see Figs. 3.15 and 3.16). The development of the aquiferous system is a species-specific pattern, which determines the resulting growth form for an exigent part.

Fig. 3.14. Diagrammatic view of the aquiferous system of the sponge *Haliclona oculata*. The water enters through the inhalant pores, the suspended material in the water is filtered away in the tissue and the water leaves the sponge through the oscula. Only close to the oscula are tracks of canals of the aquiferous system visible.

3.6 An Iterative Geometric Construction Simulating the Radiate Accretive Growth Process of a Branching Organism

In the two previous sections on the stony coral *Montastrea annularis* and the sponge *Haliclona oculata*, in the longitudinal sections and tangential views a similar structure can be observed. Although the two structures emerged in different ways in taxonomically very different organisms, and also consist of chemically very different materials, there is an essential correspondence. In both cases the surfaces of the growth layers can be simplified as a regular tessellation with pentagons and hexagons (compare Figs. 3.10B and 3.13) and the longitudinal sections display the same radiate accretive structure (compare Fig. 3.9 and Fig. 3.11B). The formation of this generic structure, found in very different taxonomical groups, is the principle of the 2D model discussed in this section. In this section the development of a general model for radiate accretive growth is discussed using the autotrophic coral *Montastrea annularis* and the heterotrophic sponge *Haliclona oculata* as examples.

The general model for radiate accretive growth will be developed stepwise (comparable with the development of the iterative geometric constructions, shown in Fig. 2.20) as shown in Fig. 3.17 in which an overview is given (see also Kaandorp 1991a). For each step the biological relevance for both examples is discussed, as well as the biological relevance in general.

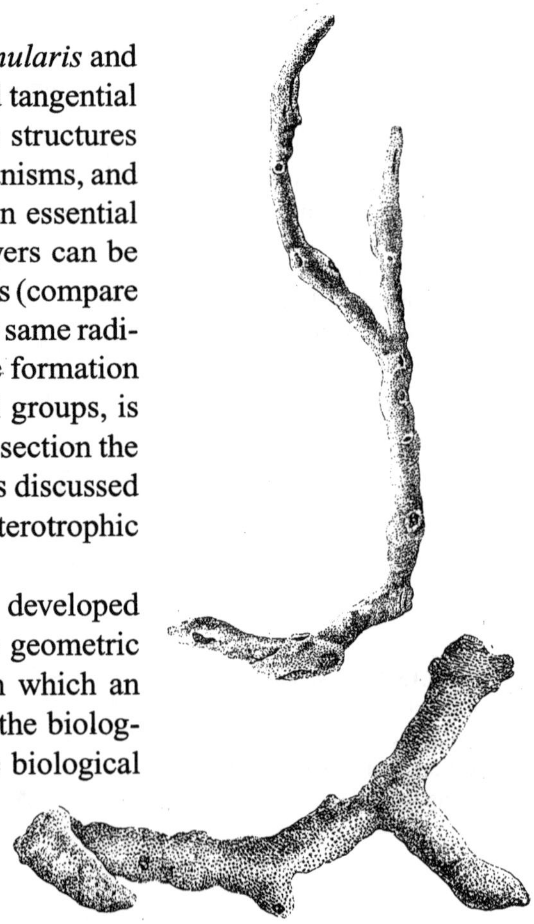

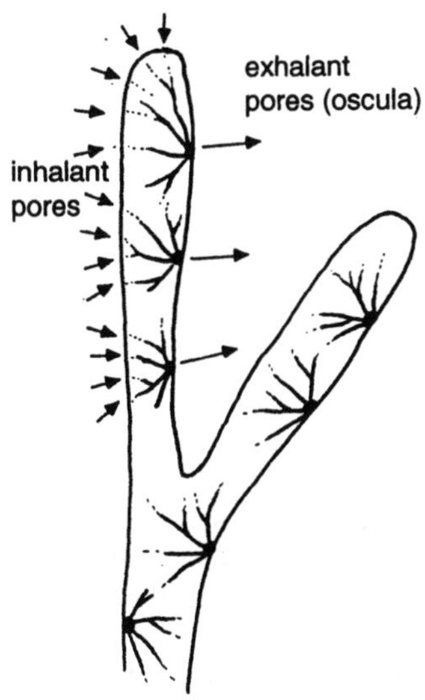

The radiate accretive growth process in two dimensions will be modelled in an iteration process (see Fig. 3.18), in which the formation of the skeleton, as seen in a longitudinal section (see Fig. 3.11B and Fig. 3.9), is simulated. In each iteration step a layer consisting of new longitudinal elements (perpendicular to the preceding layer of tangential elements) and new tangential elements (the new surface of the object) is constructed. In the 2D simulation the tangential elements are situated on growth lines of the object. This is a simplification of reality: in the actual objects the tangential elements are arranged in pentagons and hexagons (see Figs. 3.10B and 3.13). In Chap. 5 this arrangement will be discussed in detail. In Fig. 3.19 the construction of a new tangential and a new longitudinal element is shown. The length l of the longitudinal element is determined by a *generator processing function*. In the model the length of the longitudinal element varies between 0.0 and 1.0 in most cases (in one example also values between 0.0 and 2.0 are allowed), while the width s of the tangential element is a constant. After the construction of a new layer of tangential and longitudinal elements, the object should satisfy a number of rules, represented by *post-processing functions*. The *generator processing* and *post-processing functions*, which will be discussed in detail in the next subsections, are used to model the internal properties of the simulated organism as well as the influence of the physical environment on the growth process. The following subsections are subdivided into two parts: in the first part the model is discussed, and in the second part a description is given of the biological interpretation of the rules introduced in the first part. Finally in Sect. 3.9, the symbols used in this section are listed.

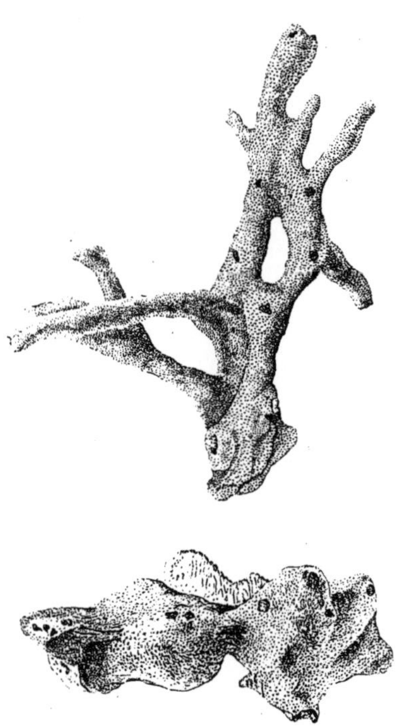

Fig. 3.15. Some growth forms of the sponge *Haliclona simulans*, closely related to *Haliclona oculata* (after Johnston, 1862)

3.6.1 The Basic Construction: the *generator*

The Model. The basic geometric construction is shown in Fig. 3.19. The *initiator* consists of a number of tangential edges with *fertilization* state "fertile", situated in a semicircle. The generator replaces each fertile edge (length s) by a non-fertile longitudinal edge and a fertile tangential edge. This geometric construction is indicated in Fig. 2.40 as the *generator*. The length of the longitudinal edge l is determined by a *generator processing function*. This function uses the original *generator* and *local_inf* as arguments. The latter is the angle α between the original tangential edge and the vertical y-axis in (3.1).

$$
\begin{aligned}
f(\alpha) &= \sin(\alpha) \text{ for } 0 \leq \alpha \leq \pi \\
l &= s \cdot f(\alpha)
\end{aligned}
\tag{3.1}
$$

Fig. 3.16. The sponge *Haliclona simulans* under natural conditions (the photograph was made near Roscoff in France by M.J. de Kluijver).

The construction can be described by a replacement system (see also Kaandorp, in press) in which the *base elements* are unconnected edges:

$$\begin{aligned} initiator &= (edge(V_0, V_1), F); \cdots (edge(V_{n-1}, V_n), F); \quad (3.2)\\ generator &= (edge(V_i, V_{i+1}), F); \rightarrow (edge(V_i, V_{i+1}), NF);\\ &\quad (edge(M_{1j}(V_i), M_{2j}(V_i)), NF);\\ &\quad (edge(M_{3j}(V_i), M_{4j}(V_i)), F);\\ &\quad (edge(V_i, V_{i+1}), NF); \rightarrow (edge(V_i, V_{i+1}), NF); \end{aligned}$$

The construction starts with an initiator consisting of n edges, for which:

$$\forall 0 \leq i < n : \| V_i, V_{i+1} \| = s \quad (3.3)$$

The transformations used in the replacement system are:

$$\begin{aligned} Dx &= 0.5 \cdot (Vx_{i+1} - Vx_i) \quad (3.4)\\ Dy &= 0.5 \cdot (Vy_{i+1} - Vy_i)\\ M_{1j} &= T(Dx, Dy)\\ &\quad \text{if } (\alpha > \pi/2) \text{ then}\\ &\quad M_{2j} = T(Dx + l \cdot \sin(\alpha), Dy - l \cdot \cos(\alpha))\\ &\quad \text{else}\\ &\quad M_{2j} = T(Dx - l \cdot \sin(\alpha), Dy + l \cdot \cos(\alpha))\\ M_{3j} &= M_{2j} \cdot T(-Dx, -Dy)\\ M_{4j} &= M_{2j} \cdot T(Dx, Dy) \end{aligned}$$

Fig. 3.17. Diagram showing the development of growth models for organisms with radiate accretive growth

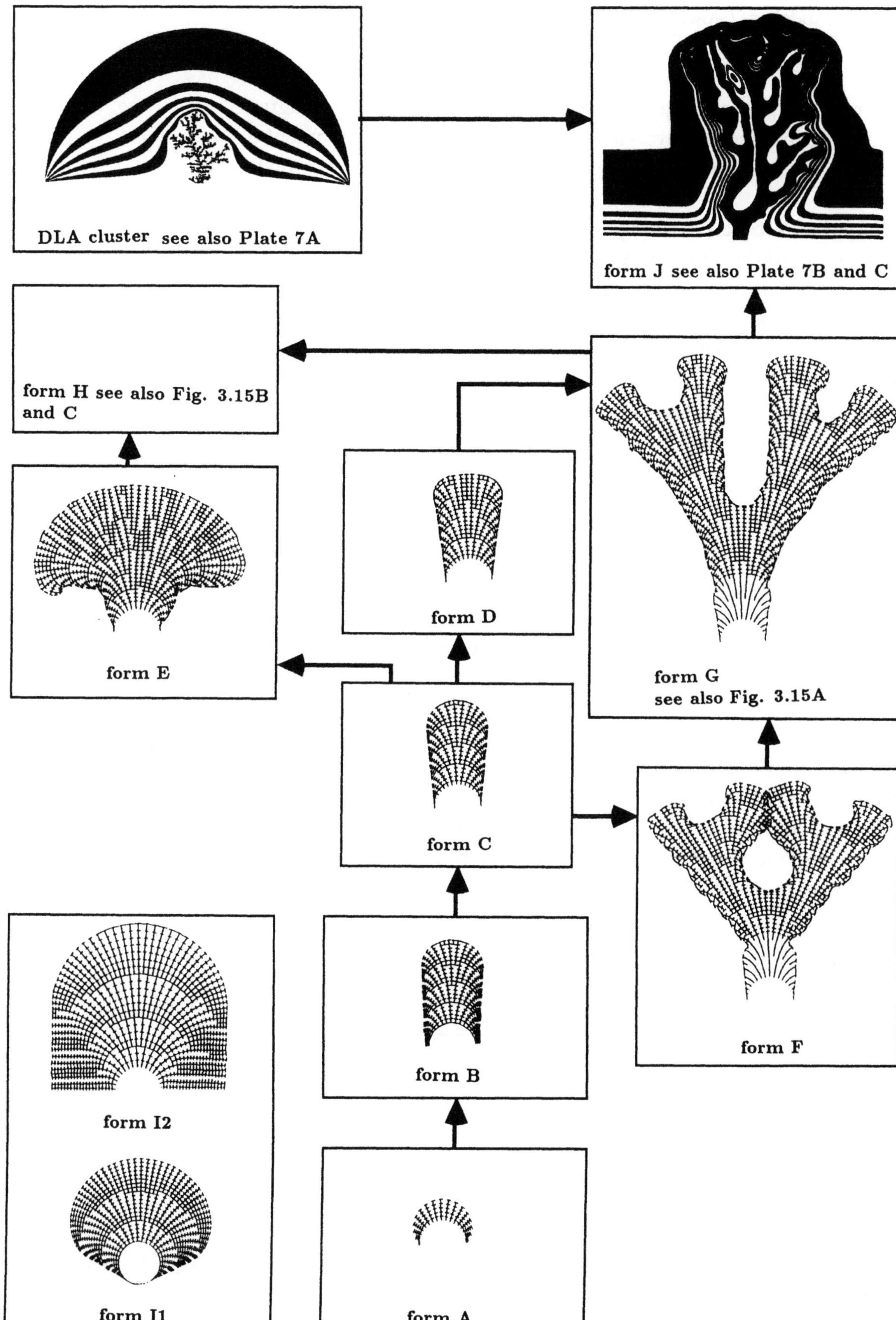

DLA cluster see also Plate 7A
form J see also Plate 7B and C
form H see also Fig. 3.15B and C
form E
form D
form G
see also Fig. 3.15A
form C
form F
form I2
form I1
form B
form A

The result of this construction appears at the bottom of the diagram in Fig. 3.17A, in which an overview is presented of the development of the growth models of a branching organism with radiate accretive growth.

Tangential view of Halicluna oculata

The Biological Objects. A tangential view of a microscopical section of *Haliclona oculata* (Fig 3.13A) shows that the size of the tangential elements scarcely varies. The tangential elements are arranged in a pattern of pentagons and hexagons, where each element consist of one spiculum. The biological interpretation of the constant s in the model is the discrete size of one spiculum, as the size of the individual spicula vary only slightly. The longitudinal element can vary in length because of the construction of the longitudinal bundles (see Fig. 3.13B), which consists of a row of about two spicula thick. The maximal length of the longitudinal bundles will depend on the size s of one spiculum; for simplicity reasons the maximal length of l is chosen to be s. In reality this will depend on species-specific characteristics in the skeleton architecture. The maximal length, for example, can easily become larger in species where the longitudinal bundles consist of rows of more than two spicula thick.

Tangential view of Montastrea annularis

In a tangential view of *Montastrea annularis* (Fig. 3.10B) and many other Scleractinia, the same arrangement of the tangential elements, in mainly pentagons and hexagons, can be observed. The size of the tangential elements can only vary in a very limited way because the stony coral colony is built of discrete modules, the theca holding the polyps, where only a small variation in size can occur. The longitudinal size of the theca, the corallites on the longitudinal section in Fig. 3.9, can vary in length, because the skeleton, consisting of calcium carbonate, is secreted at the lower base (the basal disc) of the polyp. The calcium carbonate is

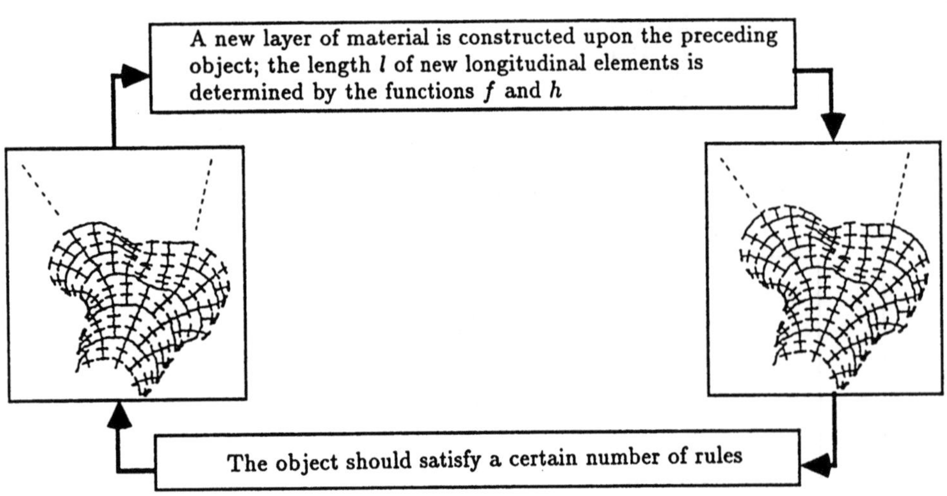

Fig. 3.18. Iteration process for modelling the radiate accretive growth process

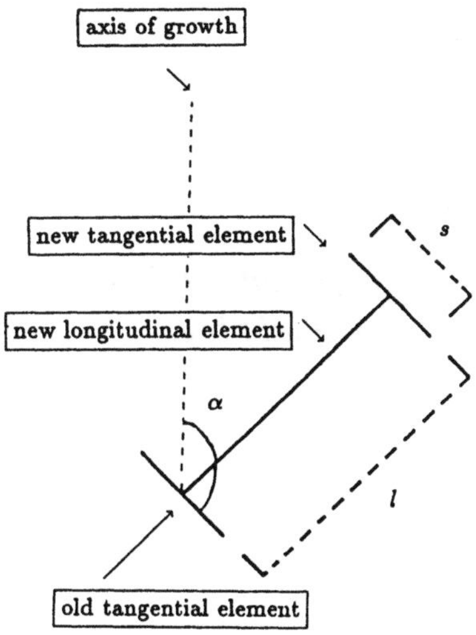

Fig. 3.19. Construction of a new tangential and a new longitudinal element. The length l of the longitudinal element is determined by a *generator processing function*, s is the length of a tangential element, and α is the angle between the axis of growth (dotted line) and the old tangential element.

deposited beneath the living tissue of the polyps upon a dead core, which makes a continuous variation in length possible. In stony corals these longitudinal lengths are influenced by local differences in deposition rates on the colony.

In general the constant s in the model reflects the observation that most organisms with radiate accretive growth are built from discrete units (modules, skeleton elements, etc.), while the growth process itself causes a continuous variation in the longitudinal elements l.

The initiator (see Fig. 3.17) used in the iteration process, in all the 2D constructions shown in this section, consists of a number of tangential edges situated in a semicircle. The form of the initiator is chosen because the growth process of many organisms with radiate accretive growth starts as an encrusting layer, where from small hemispherical protrusions the branches develop.

Longitudinal sections of *Haliclona oculata* (Fig. 3.11B and Fig. 3.12) show that the growth velocity depends upon the angle between the tangent of the surface and the axis of a column-shaped sponge tip. The highest growth velocities occur in parts of the sponge where the surface and the axis of growth make an angle of 90 degrees, while the velocities decrease to zero at an angle of about 180 degrees. This phenomenon can be explained by the presence of the spicula secreting cells, which exhibit the highest activity close to the tip, where they are situated in an area with the highest supply of material suspended in the water. Indeed, in sections of growing sponge tips, where the spicula secreting cells are stained (examples of such sections of the branching sponge species *Spongilla lacustris* can be found in Brien et al. (1973)), it can be seen that the spicula secreting activity is the highest at the tips of the sponge and decreases towards the

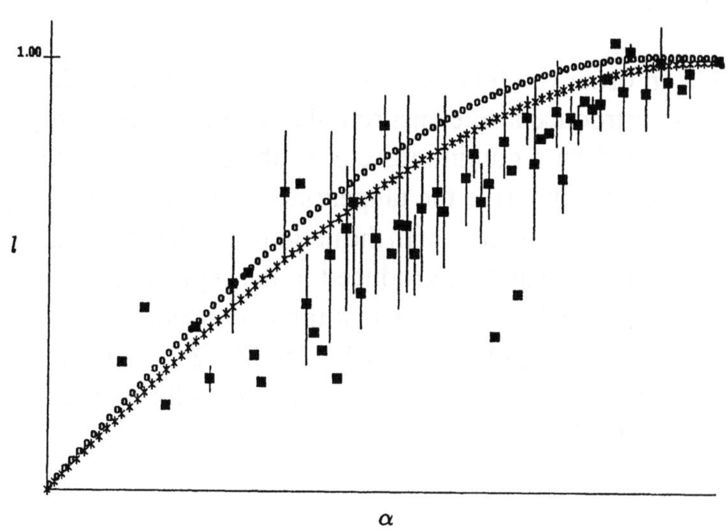

Fig. 3.20. Plot of measurements of the (normalized) length of l longitudinal fibres between two succesive growth lines (the sample from Fig. 3.11A was used) and the angle α between the longitudinal fibre and the axis of growth. The measurements are indicated with '2' marks. In the same graph two versions of $f(\alpha)$ are shown: the '$*$' marks indicate $f(\alpha)$ from (3.1) and the 'o' marks the one from (3.11).

sides. Towards the sides the access to suspended material as well as the secretion decreases. In the heterotrophic example, with internal secretion of elements, growth of the longitudinal elements is related to the angle α between a tangential element and an axis of growth. This dependence is modelled by the function $f(\alpha)$ (3.1).

Measurements of the longitudinal fibres

In Fig. 3.20, measurements of longitudinal fibres, done in the sample from Fig. 3.11A, are plotted on a graph. The length l of the longitudinal fibres between two successive growth lines and their angle α with the axis of growth were measured. In this figure the measurements were normalized using the length of the longitudinal fibre parallel to the axis of growth. The mean values are indicated with '□' marks and the minimal and maximal value for a given α displayed as a vertical line. This graph is based on 141 measurements in 8 different growth rings. On the same graph the function $f(\alpha)$ is plotted (visualized with '$*$' marks); it can be seen that this function can be used to approximate the rate of secretion of longitudinal elements as a function of the angle α.

This function $f(\alpha)$ is only relevant for organisms with radiate accretive growth with internal secretion of the elements. It is not relevant for Scleractinia as *Montastrea annularis*, since in these organisms the material is deposited externally on a dead core. For organisms with internal secretion this function $f(\alpha)$ is the basic cause of the radiate accretive structure. In the autrophic class with external secretion the radiate accretive structure has a different cause, but this subject will be discussed in more detail in the section on the modelling of the influence of the light intensity on the growth process.

External secretion

3.6.2 Modelling the Coherence of the Skeleton

The Model. In order to obtain continuous growth lines in the model it is necessary to introduce a rule ensuring that neighbouring tangential elements are situated on a continuous curve (the *continuity rule*, see also Fig. 3.17B). Without this rule the tangential elements are growing independently, without a connection with the neighbouring tangential elements (Fig. 3.17A). This rule can be introduced in the replacement system as a *post-processing* rule, which ensures that three adjacent tangential edges $((a0, a1); (M_3 j(V_i), M_4 j(V_i)); (c0, c1)$; see Fig. 3.21) are situated on a continuous curve. The replacement system in which this context sensitive rule is introduced is listed below:

Continuity rule

$$\begin{aligned}
initiator &= (edge(V_0, V_1), F); \cdots (edge(V_{n-1}, V_n), F); \quad (3.5)\\
generator &= (edge(V_i, V_{i+1}), F); \to (edge(V_i, V_{i+1}), NF);\\
&\quad (edge(M_{1j}(V_i), M_{2j}(V_i)), NF);\\
&\quad (edge(M_{3j}(V_i), M_{4j}(V_i)), F);\\
&\quad (edge(V_i, V_{i+1}), NF); \to (edge(V_i, V_{i+1}), NF);\\
continuity\,rule &= M_{3j}(V_i) \to b0,\; M_{4j}(V_i) \to b1
\end{aligned}$$

The new vertices $b0$ and $b1$ (see also Fig. 3.21) are determined in the following way:

$$\begin{aligned}
pb &= M_{2j}(V_i) \quad (3.6)\\
pa &= 0.5 \cdot (a0 + a1)\\
pc &= 0.5 \cdot (c0 + c1)\\
d &= \|pa, pc\|\\
b0 &= pb - 0.5 \cdot s/d(pa - pc)\\
b1 &= pb + 0.5 \cdot s/d(pa - pc)
\end{aligned}$$

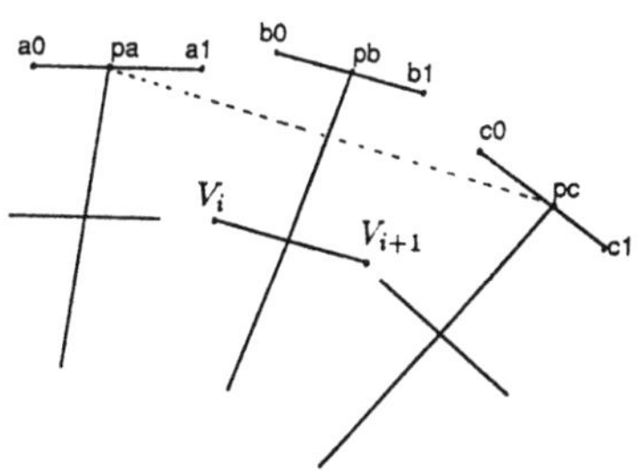

Fig. 3.21. Construction of the new vertices $b0$ and $b1$, satisfying the *continuity rule* (see (3.6) and (3.5))

The result of this rule is that the edge($b0, b1$) is set parallel to the edge(pa, pc).

A second growth rule, which is necessary to obtain continuous growth lines, is the addition of new tangential elements when there is enough space between two neighbouring elements (*insertion rule*, see also Fig. 3.17B). Without this rule, gaps would appear in the growing object, increasing in size after each iteration step; the tangential and longitudinal elements would not form a coherent skeleton any longer and the skeleton would disintegrate. This rule is introduced in the following replacement system, where a new tangential edge ($n0, n1$) is inserted between two neighbouring tangential edges ($b0, b1$) and ($c0, c1$) (see Fig. 3.22):

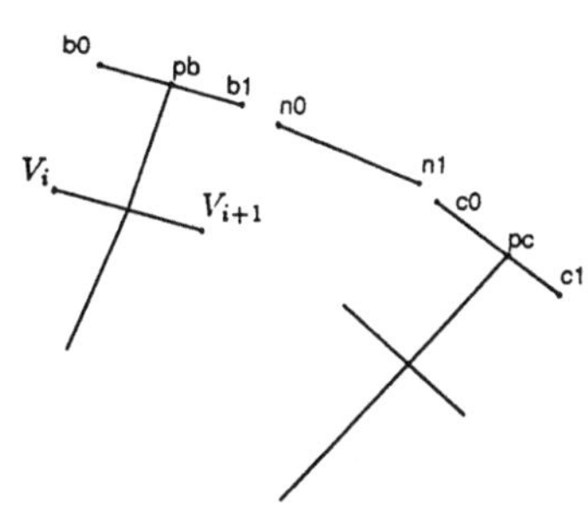

Fig. 3.22. Construction of a new tangential element with vertices $n0$ and $n1$ in the *insertion rule* (see (3.8) and (3.7))

$$\begin{aligned}
initiator &= (edge(V_0, V_1), F); \cdots (edge(V_{n-1}, V_n), F); \quad (3.7)\\
generator &= (edge(V_i, V_{i+1}), F); \to (edge(V_i, V_{i+1}), NF);\\
&\quad (edge(M_{1j}(V_i), M_{2j}(V_i)), NF);\\
&\quad (edge(M_{3j}(V_i), M_{4j}(V_i)), F);\\
&\quad (edge(V_i, V_{i+1}), NF); \to (edge(V_i, V_{i+1}), NF);\\
continuity\,rule &= M_{3j}(V_i) \to b0,\; M_{4j}(V_i) \to b1\\
insertion\,rule &= \text{if } (\|b1, c0\| > s) \text{ then}\\
&\quad (edge(b0, b1), F); \to (edge(b0, b1), F);\\
&\quad (edge(n0, n1), F);
\end{aligned}$$

The vertices $n0$ and $n1$ of the newly inserted tangential edge (see also Fig. 3.22) are determined in (3.8).

$$\begin{aligned} pb &= M_{2j}(V_i) \\ pc &= 0.5 \cdot (c0 + c1) \\ d &= \|b1, c0\| \\ n0 &= b1 + 0.5 \cdot (d - s)/d \cdot (c0 - b1) \\ n1 &= b1 + (s + 0.5 \cdot (d - s))/d \cdot (c0 - b1) \end{aligned} \tag{3.8}$$

The Biological Objects. By applying both post-processing rules, object Fig. 3.17B with a coherent skeleton and continuous growth lines can be obtained from object A. In reality tangential and longitudinal elements are arranged in continuous lines, caused by the fact that the spicula bundles form one connected mesh (see: Figs. 3.13A and B) in the example of the sponge *Haliclona oculata*. In the case of *Montastrea annularis* and other Scleractinia, the theca of the polyps cannot grow independently from their neighbours: they are imbedded in the same structure. In general this phenomenon, that units in a growing organism cannot grow independently from the neighbouring units because they are embedded in a common structure, is represented in the model by the *continuity rule*. *Continuity rule*

The second rule applied in the iteration process is also related to the conservation of the coherence of the skeleton. During the growth process the circumference of the model as well as the real organism with radiate accretive growth is increasing continuously. The consequence is that the size of the polygons tessellating the surface of the organism (see for example *Haliclona oculata* Fig. 3.13A and *Montastrea annularis* Fig. 3.10B) would increase in subsequent growth steps, when no new tangential elements are inserted. In reality the tangential elements in *Haliclona oculata*, with a size of one spiculum, cannot remain connected and the skeleton would collapse. In the case of stony coral, gaps would occur between the theca of neighbouring polyps. In order to conserve coherence new polyps have to be inserted (compare Fig. 2.6). The insertion of new tangential elements to preserve the coherence of the skeleton is simulated by the *insertion rule*. The insertion of new tangential elements causes the emergence of new longitudinal lines; this can also be observed in both longitudinal sections (see arrows in Fig. 3.11B and Fig. 3.9). Both coherence conserving rules, the *continuity* and the *insertion rule*, are relevant for all types of organisms with radiate accretive growth. *Insertion rule*

3.6.3 Introduction of the Smallest Skeleton Element in the Model

The Model. To generate columns which do not accumulate material at the lateral sides (Fig. 3.17B) it is necessary to introduce a new rule in the *generator processing function*: when l drops below a certain $inhibition\_level$, growth will stop. In (3.9) this stopping rule is included.

$$
\begin{aligned}
f(\alpha) &= \sin(\alpha) \text{ for } 0 \leq \alpha \leq \pi \\
l &= \begin{cases} s \cdot f(\alpha) \text{ for } f(\alpha) > inhibition\_level \\ 0.0 \text{ for } f(\alpha) \leq inhibition\_level \end{cases}
\end{aligned}
\tag{3.9}
$$

Fertilization attribute

The result is shown in Fig. 3.17C. When growth of a fertile edge is inhibited, the *fertilization* state of this edge is changed into "non-active". The growing object is limited by active and non-active fertile edges; only active edges will participate normally in generating new edges during subsequent growth stages. In this object growth only takes places at the tip. It is useful to make a distinction between active and non-active fertile edges. With this distinction it is possible that non-active tangential elements, under certain circumstances, can participate again in the growth process. Their state is then changed into active, and growth (which will be indicated as secondary growth) can take place. In the replacement system the edges which are in the state "non-active" are indicated with the attribute 'SF'. The *generator* part of this replacement system is displayed below:

Secondary growth

(3.10)

$$
\begin{aligned}
generator \;=\;& (edge(V_i, V_{i+1}), F); \rightarrow (edge(V_i, V_{i+1}), NF); \\
& (edge(M_{1j}(V_i), M_{2j}(V_i)), NF); \\
& \text{if } (l > inhibition\_level) \text{ then} \\
& \quad (edge(M_{3j}(V_i), M_{4j}(V_i)), F); \\
& \text{else} \\
& \quad (edge(M_{3j}(V_i), M_{4j}(V_i)), SF); \\
& (edge(V_i, V_{i+1}), NF); \rightarrow (edge(V_i, V_{i+1}), NF); \\
& (edge(V_i, V_{i+1}), SF); \rightarrow (edge(V_i, V_{i+1}), SF);
\end{aligned}
$$

The Biological Objects. The biological interpretation of the threshold value $inhibition\_level$ in (3.9) is the minimal possible length of the longitudinal elements. The longitudinal element in *Haliclona oculata* can vary in length because of the construction of the longitudinal bundles (see Fig. 3.13B), which consists of a row of about two spicula thick. The length l can vary between a certain upper limit, which is species-specific and depends on s, the size of a tangential element, and a certain lower limit $inhibition\_level$. The interpretation of the lower limit $inhibition\_level$ is that the skeleton is built of discrete elements and the longitudinal bundles cannot become arbitrarily short. This would lead to a very dense skeleton where there is no space left for the living tissue. In reality this accumulation is not found and elements do not develop in the skeleton when there is not enough space. In the sponge *Haliclona oculata* (Fig. 3.11B) growth stops at the lateral sides of the object when the length of the longitudinal edge drops below a certain threshold. This growth stop at the lateral sides can also be observed in Fig. 3.12. Here no material is added during the growth process over the horizontally situated needles. The stopping of growth when the size of new elements drops below a certain threshold is expressed in the modified generator processing function in (3.9). Under certain circumstances, for example when the tissue of the sponge is damaged locally, secondary growth can take place at these non-active sites (in the section on the transplantation experiments with *Haliclona oculata* some examples will be shown). In the model this can be simulated by changing the status of non-active fertile edges again into active.

Minimum size longitudinal elements in Haliclona oculata

Secondary growth

The lower limit for the longitudinal elements is only relevant for organisms with radiate accretive growth with internal secretion. In the Scleractinia where the material is secreted superficially the length of the longitudinal elements can be vary between almost zero and the upper limit. A growth velocity of zero in the Scleractinia indicates local dying of polyps. In the longitudinal section of *Montastrea annularis* (Fig. 3.9) it can be observed that in the lateral corallites, which make the largest angle with the axis of growth of the colony, still a small amount of material is added to the colony (in Graus and Macintyre 1982, this effect is indicated as the "maximum corallite growth angle with respect to the vertical").

Minimum size longitudinal elements in Montastrea annularis

3.6.4 Modelling the "Widening Effect"

The Model. The column shown in Fig. 3.17C can be flattened by using a generator processing function in which an area of equal values appears instead of one maximum:

$$f(\alpha) = \begin{cases} 1.0 \text{ for } \pi/2 - \pi/w \leq \alpha \leq (\pi/2 + \pi/w) \\ \sin((\pi/2)/(\pi/2 - \pi/w) \cdot (\pi - \alpha)) \text{ for} \\ 0 \leq \alpha < (\pi/2 - \pi/w), (\pi/2 + \pi/w) > \alpha \leq \pi \end{cases} \tag{3.11}$$

$$l = \begin{cases} s \cdot f(\alpha) \text{ for } f(\alpha) > inhibition\_level \\ 0.0 \text{ for } f(\alpha) \leq inhibition\_level \end{cases}$$

$$w > 2$$

A more or less flattened shape is obtained by choosing different values for w (see Fig. 3.17D).

Widening factor

The Biological Objects. The constant w represents a widening factor in function f; the biological relevance is clear from the observation that in the growth process of a branching organism with radiate accretive growth the tip widens before it splits into new branches (see for example the longitudinal sections of *Haliclona oculata* in Fig. 3.11B and Fig. 3.12). Without widening, the sponge would remain a non-branching column, as will be explained in the section on the formation of branches. The widening represents a region in the tip where the secreting activity is not influenced by a lower access to the suspended material in the environment. This causes a small area of equal maximal growth velocities at the tip of an organism with radiate accretive growth. The widening factor is a species-specific parameter in the model. The function $f(\alpha)$ is an approximation of the secretion of longitudinal elements as a function of α. In Fig. 3.20 this function (indicated with 'o' marks) is plotted together with measurements (indicated with '□' marks) done on the length of the actual longitudinal fibres.

3.6.5 Formation of New Growth Axes

The Model. So far, growth rules were formulated for columnar forms. These forms have in common that there is one growth-axis, the y-axis. In order to create branching forms it is necessary to formulate a new rule, allowing the generation of new growth axes. These might arise on the surface of the growing object on sites where a (local) maximum in growth velocities occurs. The emergence of these maxima will be discussed in more detail in the next subsections. In Fig. 3.23A a growing object is shown in which two local maxima develop. The first seven layers in the object are fertile tangential elements associated with a single growth axis ($prev\_DA$). In the 8th layer two local maxima and one minimum develop. In the 8th and next layers, fertile elements are associated with new growth axes new_DA_0 (the fertile elements to the left of the local minimum) and new_DA_1 (the elements to the right of the local minimum). The association of fertile elements with a specific growth axis is determined in a new post-processing rule (the *association rule*). The angle α between a growth axis and a fertile tangential element is calculated in the *generator processing function*. The edge replacement system is extended with a new attribute which defines the direction of the growth axis (the vector $[DA_x, DA_y]$). In the replacement system, (DA_x, DA_y) is initially set to

Association rule

(0,1). The direction of new growth axes corresponds to the direction of longitudinal elements, where local maxima develop. The fertile tangential elements are associated with the nearest growth axis in a procedure as described in Fig. 3.23A. In the *association rule* the previous growth axis $(prev\_DA_x, prev\_DA_y)$ is replaced by the direction of the new formed growth axis (new_DA_x, new_DA_y).

Direction new growth axes

The algorithm in which fertile elements are associated with a specific growth axis is described in pseudo code below. The algorithm consists of two parts. In part *A* the length of longitudinal elements connected with tangential elements situated at the surface of the object are compared to each other. The longitudinal elements with a minimum maximum value are stored in two separate lists. In part *B* the tangential elements are associated with a growth axis.

Algorithm association rule

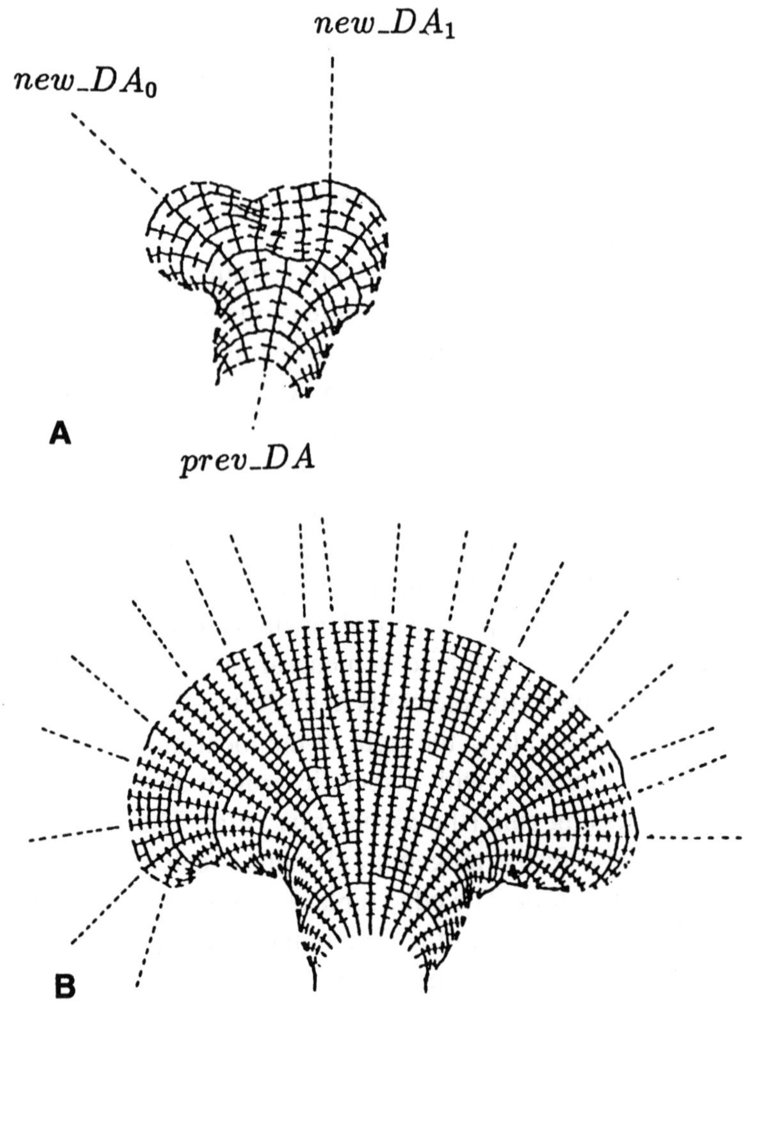

Fig. 3.23. Development of new growth axes in the model. In example A an old axis of growth $prev\_DA$ is replaced by two new ones (new_DA_0 and new_DA_1), in example B many new axes of growth are generated due to a slight disturbance of the growth process.

```
association_rule( objects ) {
  initialization list of maxima, list of minima, list of plateau values;
  initialization booleans ascending, plateau;
  initialization variables containing the length of a longitudinal element
  cur_l, prev_l;
  A: next tangential element in the state F or SF is taken from the objects {
    length cur_l of the longitudinal element connected with the current
    tangential element is determined;
    if (cur_l > prev_l) {
      if (ascending == FALSE ) {
        if (plateau == TRUE) {
          the middle element of the list of plateau values is added to the
          list of minima; the list of plateau values is re-initialized; }
        else current longitudinal element is added to the list of minima;
      ascending = TRUE; plateau = FALSE; } }
    else if (cur_l < prev_l) {
      if (ascending == TRUE ) {
        if (plateau == TRUE) {
          the middle element of the list of plateau values is added to the
          list of maxima; the list of plateau values is re-initialized; }
        else current longitudinal element is added to the list of maxima;
      ascending = FALSE; plateau = FALSE; } }
    else {
      current longitudinal element is added to the list of plateau values;
      plateau = TRUE; }
    prev_l = cur_l;
  } end A
  first maximum is taken from the list of maxima;
  first minimum is taken from the list of minima;
  B: next tangential element in the state F or SF is taken from the objects {
    the current tangential element is associated with the growth axis,
    the growth axis (the vector [DA_x, DA_y]) corresponds with the direction
    of the longitudinal element in the current maximum;
    if (longitudinal element connected with the current tangential element
    corresponds to the current minimum) {
      next maximum is taken from the list of maxima;
      next minimum is taken from the list of minima; }
  } end B
} end association_rule
```

The Biological Objects. The biological meaning is that those parts of the real organism with radiate accretive growth are situated in areas with the strongest water movement, or under sheltered conditions in areas which are not yet depleted in nutrients by the other branches, where they will have easy access to the suspended food and will develop the highest growth velocities. As a consequence, the protrusion increases in size and new branches (axes of growth) are formed.

Access to suspended food

3.6.6 Disturbance of the Growth Process, Formation of Plates

The Model. The growth function described in (3.9) will never generate more than one maximum. More maxima can emerge when the growth process is disturbed by external influences. A simple example is the superposition of "noise" on the final length l of the longitudinal edge. In the generator processing function (3.12) a function g is introduced, which returns random values between two limits $lowest\_value$ and 1.0.

$$f(\alpha) = \sin(\alpha) \text{ for } 0 \leq \alpha \leq \pi \tag{3.12}$$

$$g(lowest\_value) = random\_function(lowest\_value, 1.0)$$
$$for\ 0.0 \leq lowest\_value \leq 1.0$$

$$l = \begin{cases} s \cdot f(\alpha) \cdot g(lowest\_value) \text{ for} \\ f(\alpha) \cdot g(lowest\_value) > inhibition\_level \\ 0.0 \text{ for} \\ f(\alpha) \cdot g(lowest\_value) \leq inhibition\_level \end{cases}$$

In (3.12), l is determined by the product of the function $f(\alpha)$ (from (3.9) and the function $g(lowest\_value)$. Even for a slight disturbance (a value for $lowest\_value$ just below 1.0) the form is deregulated and plate-like forms as shown in Fig. 3.17E are generated. In those plate-like forms new irregularities are generated during each iteration, which produce new growth axes in turn. This process is depicted in Fig. 3.23B.

Plate-like forms

The Biological Objects. In reality the growth process will be more disturbed on exposed sites; in a turbulent environment a large variation in growth velocities will occur. Protrusions might also occur as irregularities on the surface. Because of this variation local maxima and minima in growth velocity can be identified on the growing object. The most protruding parts of the sponge will have the highest access to material suspended in the water and will develop the highest growth velocities. A turbulent environment is modelled by the superposition of some "noise" on the final length of l in (3.17). The degree of disturbance is controlled with the parameter $lowest\_value$.

Controlling the degree of disturbance

3.6.7 Additional Rules for the Formation of Branches and Plates

The Model. Two new post-processing rules are necessary when new growth axes are to be generated during the formation of plates and branches. During this process some active fertile edges will be enclosed by surrounding active fertile edges. In order to prevent collisions, edges of this type are removed from the object. After this, longitudinal lines so far always ending in a tangential edge will now end somewhere in the object. This post-processing rule (*deletion rule*) is the reverse of the *insertion rule*. Without the removal of non-fitting elements from the skeleton a dense accumulation of elements would appear. In the following replacement system two new post-processing rules are included:

Deletion rule

$$
\begin{aligned}
initiator \;&=\; (edge(V_0, V_1), F(0, 1)); \cdots \\
&\quad\; (edge(V_{n-1}, V_n), F(0, 1)); \\
generator \;&=\; (edge(V_i, V_{i+1}), F, (prev\_DA_x, prev\_DA_y)); \rightarrow \\
&\quad\; (edge(V_i, V_{i+1}), NF); \\
&\quad\; (edge(M_{1j}(V_i), M_{2j}(V_i)), NF); \\
&\quad\; \text{if } (l > inhibition\_level) \text{ then} \\
&\quad\quad (edge(M_{3j}(V_i), M_{4j}(V_i)), F, (prev\_DA_x, prev\_DA_y)); \\
&\quad\; \text{else} \\
&\quad\quad (edge(M_{3j}(V_i), M_{4j}(V_i)), SF, (prev\_DA_x, prev\_DA_y)); \\
&\quad\; (edge(V_i, V_{i+1}), NF); \rightarrow (edge(V_i, V_{i+1}), NF); \\
&\quad\; (edge(V_i, V_{i+1}), SF, (prev\_DA_x, prev\_DA_y)); \rightarrow \\
&\quad\; (edge(V_i, V_{i+1}), SF, (prev\_DA_x, prev\_DA_y)); \\
continuity\ rule \;&=\; M_{3j}(V_i) \rightarrow b0, M_{4j}(V_i) \rightarrow b1 \\
insertion\ rule \;&=\; \text{if } (\|b1, c0\| > s) \text{ then} \\
&\quad\quad (edge(b0, b1), F, (prev\_DA_x, prev\_DA_y)); \rightarrow \\
&\quad\quad (edge(b0, b1), F, (prev\_DA_x, prev\_DA_y)); \\
&\quad\quad (edge(n0, n1), F, (prev\_DA_x, prev\_DA_y));
\end{aligned}
\tag{3.13}
$$

$$
\begin{aligned}
\textit{deletion rule} &= \text{if } (\|b1, c0\| < s) \text{ then} \\
&\quad (edge(b0, b1), F, (prev\_DA_x, prev\_DA_y)); \rightarrow; \\
\textit{association} & \\
\textit{rule} &= prev\_DA_x \rightarrow new\_DA_x, prev\_DA_y \rightarrow new\_DA_y \\
\textit{anti}-\textit{collision} & \\
\textit{rules}: & \\
1)\ \textit{non}-\textit{intersection} & \\
\textit{rule} &= \text{if } (edge(b0, b1) \text{ intersects another edge object }) \text{ then} \\
&\quad (edge(b0, b1), F, (new\_DA_x, new\_DA_y)) \rightarrow \\
&\quad (edge(b0, b1), SF, (new\_DA_x, new\_DA_y)) \\
2)\ \textit{avoidance} & \\
\textit{rule} &= \text{if } (\|pb, pv\| < s) \text{ then} \\
&\quad (edge(b0, b1), F, (new\_DA_x, new\_DA_y)) \rightarrow \\
&\quad (edge(b0, b1), SF, (new\_DA_x, new\_DA_y))
\end{aligned}
$$

Another colliding situation arises when plates or branches intersect. A rule preventing intersections (*non-intersection rule*) changes the status of intersecting active fertile elements into non-active. A slightly different version of the previous rule is a rule in which branches are not allowed at all to intersect, but are forced to keep a certain distance (*avoidance rule*). This rule changes the status of active fertile elements, that are approaching too much, into non-active.

Non-intersection rule

Avoidance rule

The Biological Objects. Without applying one of the two "anti-collision" rules, a physically impossible situation would emerge where branches of the object would grow through each other. In the replacement system of (3.13) both anti-collision rules are described. In this replacement system the vertex pv is a midpoint on a tangential edge, which is not one of the neighbouring tangential edges from Fig. 3.21.

The deletion of new tangential elements causes the disappearance of longitudinal fibres in the growth process. This can also be observed in the longitudinal section (see arrow 2 in Fig. 3.11B). Without the removal of non-fitting elements from the skeleton a dense accumulation of elements would appear and the coherence of skeleton would be disturbed. This *deletion rule* is another coherence conserving rule; together with the *continuity* and the *insertion rule*, it is relevant for all types of organisms with radiate accretive growth.

In the model, growth ceases as soon as branches are at the point of intersection. This *non-intersection rule* is, applied in a 2D model, a simplification of reality. In actual branching objects (see Fig. 3.3) branches that are at the point of intersection can escape each other in the third dimension. In many marine sessile organisms, branches which intersect also fuse with each other (anastomosis). This can also be observed in Fig. 3.3 and in branching hydrocorals (see Fig. 3.8). Both aspects will be discussed further in the section on the restrictions of the 2D model.

Anastomosis

Natural collision detection

Under natural conditions, especially on a sheltered growth site, water movement and food supply as well as the growth velocity will be slowed down or stopped when branches collide. Fig. 3.12 reveals that growth stops in the right branch as soon as the (marked) left branch overgrows the right branch. In many stony corals it can be observed that when branches approach each other too closely growth is suppressed (this is well described for the stony coral *Acropora formosa* in Oliver 1984). In some stony corals (such as the genus *Galaxea fascicularis*, with an equidistant gap between the theca) this can even be observed in neighbouring polyps. There is often a well-defined, species-specific distance between approaching polyps. This aspect of the growth process, where branches are not allowed to approach each other too closely, can be modelled with the *avoidance rule*. The effect that the growth velocity decreases because approaching branches locally deplete the same nutrient source can be modelled in a more natural way by including the influence of local nutrient concentrations on the growth process. This last point is discussed in more detail in the section on the concentration gradient model.

3.6.8 Formation of Branches

The Model. In the growth rules discussed above, the longitudinal length l of newly added elements only depends on the angle α between a growth axis and the active fertile element. An alternative growth rule, in which instead of α the radius of curvature determines the longitudinal length l, is shown in (3.14). The radius of curvature can be defined as the radius of the circle through three points on the circumference of the growing object that are situated on neighbouring tangential elements. The minimum distance between these points is the length of a tangential edge s. The rad_curv becomes infinitely large when the three points are situated on a line. In the case the points are situated on a hollow site of the contour of the object rad_curv is negative. The length of the longitudinal edge becomes zero as soon as the radius of curvature is larger than a certain (fixed) maximum (max_curv). This happens, for example, when three points are situated on one line at the lateral side of a column. l attains the maximum value when the radius of curvature is less then a fixed minimum (min_curv).

Estimation of the radius of curvature

$$
\begin{aligned}
h(rad\_curv) &= 1.0 - \qquad (3.14)\\
&\quad (rad\_curv - min\_curv)/(max\_curv - min\_curv)\\
&\quad \text{for } min\_curv \leq rad\_curv \leq max\_curv\\
h(rad\_curv) &= 1.0 \text{ for } rad\_curv < min\_curv\\
h(rad\_curv) &= 0.0 \text{ for } rad\_curv > max\_curv
\end{aligned}
$$

$$
l = \begin{cases} s \cdot h(rad\_curv) \text{ for} \\ h(rad\_curv) > inhibition\_level \\ 0.0 \text{ for } h(rad\_curv) \leq inhibition\_level \end{cases}
$$

The radius of curvature is used as an argument ($local\_inf$) for the *generator processing function*, shown in (3.14). The result of this construction is shown in Fig. 3.17F. The object starts growing as a column and the longitudinal edges added at the top (close to the growth axis) are equal-sized. After some growth stages the top flattens and the curvature exceeds max_curv. When rad_curv exceeds this maximum value the value of l decreases and the result is that the object starts branching and an old axis of growth is replaced by new axes (see Fig. 3.23A). With the parameter rad_curv the amount of contact of the elements with the environment is expressed. In Fig. 3.24 the relation between rad_curv and the amount of contact with the environment is depicted. The amount of contact with the environment can be described as a quotient of the surface ΔS through which nutrient can pass and the area (with distance d from the surface) being supplied with nutrient:

Amount of contact with the environment

$$
\frac{\text{surface}}{\text{supplied area}} = \frac{\Delta S}{\Delta S.d(1 - \frac{d}{2.rad\_curv})} = \frac{2.rad\_curv}{d(2.rad\_curv - d)} \qquad (3.15)
$$

In (3.15) it can be seen that the amount of contact decreases for an increasing value of rad_curv; for $rad\_curv >> d$ this can be approximated with $h(rad\_curv)$ (1.0 is maximal contact and 0.0 minimal). A relative large value of max_curv leads to more flattened branches where some of the elements in the object are relatively far away from the environment, while a small value causes more thin-branching objects where the elements within the objects are situated closer to the environment (both effects will demonstrated in the next section). The parameter min_curv defines the lower limit: values for rad_curv below this limit do not affect the growth process. Both parameters are expressed in units s, the width of the tangential elements.

Thin-branching objects

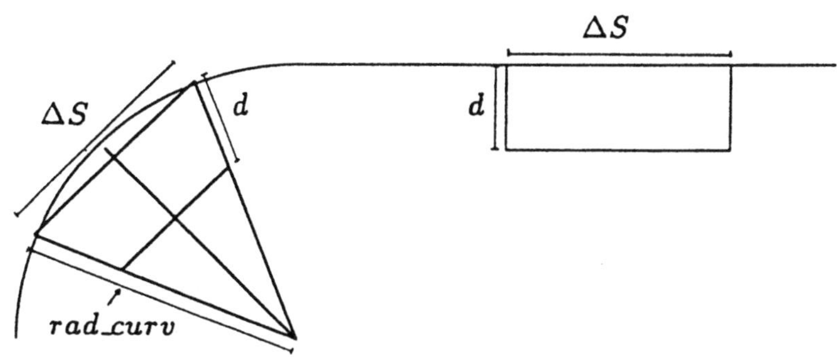

Fig. 3.24. Relation between *rad_curv* and the amount of contact with the environment: ΔS indicates the surface through which nutrient can pass to an area with distance d from the surface.

Transport of nutrients

The Biological Objects. In reality plate-like growth forms of *Haliclona oculata* (Fig. 3.17E) are found, but not under all circumstances (see Fig. 3.3). The transport of water with suspended material inside the organism is sustained only in a limited way by an aquiferous system (see Fig. 3.14) and is in *Haliclona oculata* strongly supported by external water movements as well. Under conditions with strong water movements plate-like growth forms are possible, whereas under sheltered conditions a decrease of food supply will appear in the tissue, unless it is in short-distance contact with the environment; a decrease in growth velocity results. This process is modelled with the *generator processing function* in (3.17). In this function the amount of contact with the environment is taken to be proportional to the radius of curvature. As soon as the top widens too much, the amount of contact with the environment becomes suboptimal and the growth velocity at the top decreases, resulting in a branching object.

Tip-splitting in radiate accretive growth

In general the tip-splitting for an organism with radiate accretive growth can be modelled with the function $h(rad\_curv)$ (3.14), which expresses the limitations of a certain transport mechanism of nutrients through the tissue. The parameter max_curv is a species-specific parameter which describes the power of the transport system, and depends on external water movements as well: in an environment with a higher rate of water movement, the transport system is supported by these external movements and a relatively higher value is applied for max_curv. In *Haliclona simulans* (see Fig. 3.15) where erect growth forms emerge in the same radiate accretive growth process, there is an extensive aquiferous system. A consequence is that in *Haliclona simulans* more voluminous and more globular forms develop when compared to *Haliclona oculata*. This effect can be modelled in the function h by selecting a higher value for max_curv, for a simulation of the growth process under sheltered conditions.

3.6.9 A Combination of the Previous Models

The Model. A combination of (3.11) and (3.12) yields a growth rule in which both the radius of curvature and the angle of a fertile active element with the growth axis are included. The combined growth rule can be written as a product of the growth functions: function $f(\alpha)$ is derived from (3.11) and function $h(rad\_curv)$ from (3.12). In this case $local\_inf$ in the *generator processing function* consists of two components: the rad_curv and α.

$$l = \begin{cases} s \cdot f(\alpha) \cdot h(rad\_curv) \\ \text{for } f(\alpha) \cdot h(rad\_curv) > inhibition\_level \\ 0.0 \text{ for } f(\alpha) \cdot h(rad\_curv) \leq inhibition\_level \end{cases} \tag{3.16}$$

In the growing object small areas with equal-sized longitudinal edges arise at the top. The radius of curvature increases and the value returned by h decreases. The result is that l decreases and the object starts branching (see Fig. 3.17G).

Thin-branching forms

Both effects are included in (3.16): with this *generator processing function* quite regular objects are generated. A more evolved object generated with this construction is shown in Fig. 3.25A. This object is constructed with the value $w = 18$ in $f(\alpha)$ in (3.11) and obtained after 220 iterations. In this figure only the tangential lines, the growth lines, are shown.

The next object (see Fig. 3.25B) is generated by disturbing the growth process by multiplying the product $f(\alpha).h(rad\_curv)$ ($w = 18$ in $f(\alpha)$, see (3.11)) with the function $g(lowest\_value)$ (see (3.12)). This *generator processing function* is displayed in (3.17).

$$l = \begin{cases} s \cdot f(\alpha) \cdot h(rad\_curv) \cdot g(lowest\_value) \text{ for} \\ f(\alpha) \cdot h(rad\_curv) \cdot g(lowest\_value) > inhibition\_level \\ 0.0 \text{ for} \\ f(\alpha) \cdot h(rad\_curv) \cdot g(lowest\_value) \leq inhibition\_level \end{cases} \tag{3.17}$$

Plate-like forms

The result is that the object formes plate-like branches (like Fig. 3.17E). As soon as the radius of curvature of the circumference of the plates exceeds max_curv in (3.17) the objects start branching. In this object large plates are formed by allowing a larger value for max_curv than applied in Fig. 3.25A. The plate formation at the extremities is still further increased in the object displayed in Fig. 3.25C, where the value of max_curv as well as the disturbance of the growth process (a lower value for $lowest\_value$ than in Fig. 3.25B) was increased.

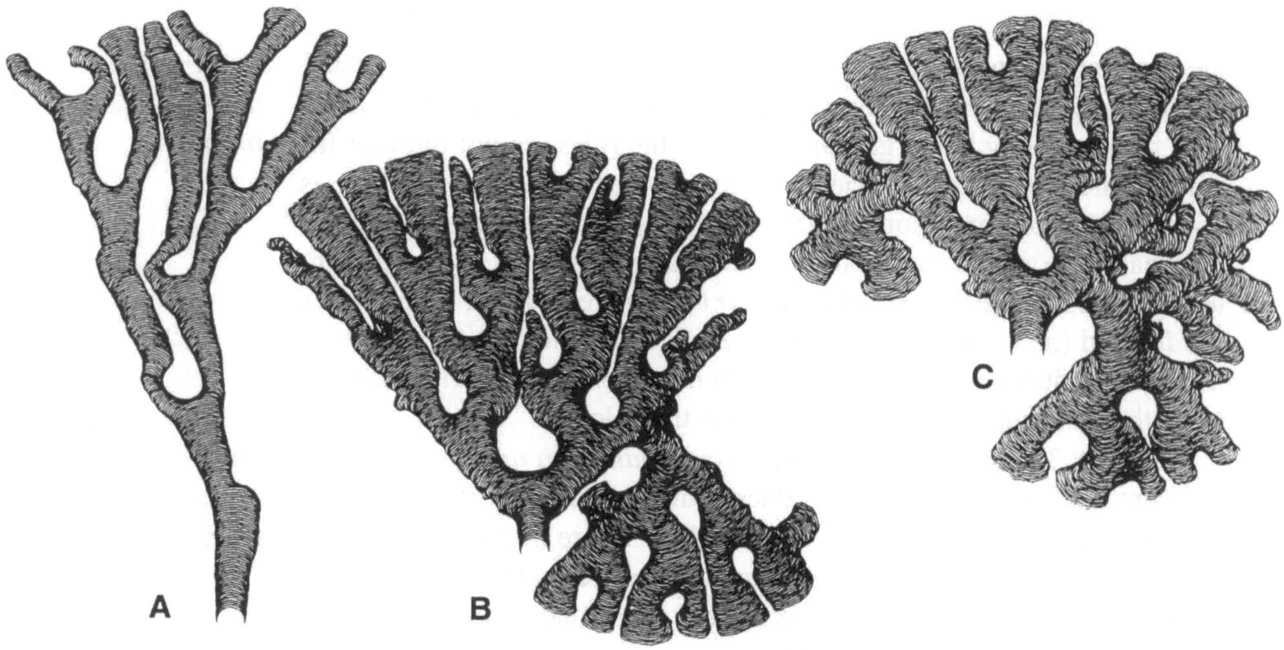

Fig. 3.25. Object A was generated with the model of Fig. 3.17G, in the objects B and C the model of of Fig. 3.17H was used. In A the *generator processing function* shown in (3.16) was used, while B and C were constructed with (3.17). In all objects the parameter n in $f(\alpha)$ (3.11) was set to the value 18. The parameter max_curv in $h(rad\_curv)$ (3.14) was respectively set to the values $7.2s$, $21.7s$, $36.1s$. In the function $g()$ (3.12) the values 0.9, 0.7 were used for $lowest\_value$ in the objects B and C respectively.

The Biological Objects. In Fig. 3.17G the model of a sponge with the highest growth velocities at the protrusions (model Fig. 3.17C) is united with the restrictions of the aquiferous system architecture (model Fig. 3.17F). The parameter max_curv in (3.14) represents the maximum allowed radius of curvature of the surface; its biological interpretation is the minimum amount of contact allowed with the environment before growth velocity will decrease and the sponge starts branching. This parameter is closely related to the degree of exposure to water movement, expressed in the model as the degree of disturbance, when $lowest\_value$ in $g(lowest\_value)$ (3.12) is increasing the parameter max_curv (3.14) can also be increased. The increase of food supply is modelled by allowing a higher maximum radius of curvature (max_curv). In the sequence Figs. 3.25A, 3.25B, 3.25C, the objects transform from thin-branching into plate-like, more irregular, and more compact forms and exhibit a higher degree of colliding branches. The same type of transformation can be seen in the range of *Haliclona oculata* (Fig. 3.3), where the thin-branching forms gradually transform in more irregular plate-like growth forms, when the exposure to water movement increases.

3.7 A Model of the Physical Environment

In the first of the following subsections an example is given of how the influence of light intensity on the growth process can be modelled, with

Autotrophic non-branching organisms

the autotrophic species *Montastrea annularis* used as example. As will be demonstrated, only a subset of the rules discussed so far are necessary for modelling this organism. Many of the rules are not relevant for autotrophic stony coral species with radiate accretive growth: there is no internal secretion of longitudinal elements, but external deposition of material upon a dead core ($f(\alpha)$ is unnecessary), there is no limiting internal transport mechanism ($h(rad\_curv)$ is unnecessary), no branches are formed (*association*, *deletion* and *avoidance* rules are superfluous). The main influence of the physical environment on the growth is assumed to be the distribution of light intensity on the colony.

Heterotrophic branching organisms

In Sect. 3.7.2 the heterotrophic species *Haliclona oculata* is used as an example. For this species the influence of the light intensity is assumed to be irrelevant. All other rules discussed so far are necessary to model the growth of this branching organism with radiate accretive growth. For a more complete model it is also necessary to include a model of the nutrient distribution around the organism.

3.7.1 The Light Model

The Model. A simple light model (Foley et al. 1990) is presented in (3.18). In this model the light intensity I (watt/m^2) on a surface is determined by $\cos(\theta)$, where θ is the angle of incidence of the light beam to the surface normal and the intensity I_S of the light source (see Fig. 3.26). The light beam corresponds to the vertical, for this tropical species the light source is assumed to be positioned in the zenith.

$$I = I_S \cdot \cos(\theta) \tag{3.18}$$

The light model can be extended by including diffuse reflection from the environment. There is reflection from the bottom as well as the surrounding water, due to scattering (see Roos 1967). If the reflected light is also included, the angle between the normal to a surface and the vertical, where I becomes nearly zero (max_angle), may vary between $\pi/2$ and π, as shown in (3.19). With the parameter max_angle the contribution of the diffuse reflection to the total light intensity can be controlled.

$$\begin{aligned} \pi/2 \le\ & max\_angle \le \pi \\ \theta^* &= \theta/max\_angle \cdot \pi/2 \\ L(\theta) &= \cos(\theta^*) \end{aligned} \tag{3.19}$$

In Fig. 3.17I two objects, I_1 and I_2, are simulated in which the length of the longitudinal elements l is determined by the the relative decrease

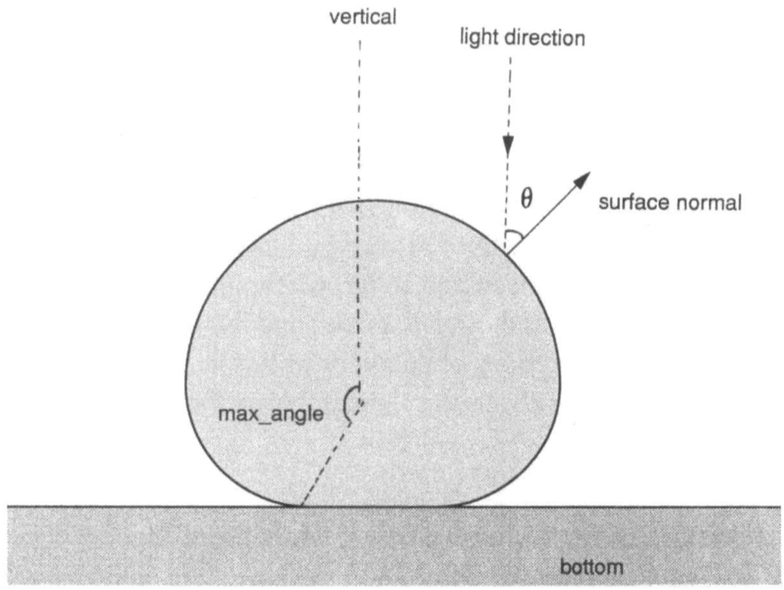

Fig. 3.26. The light intensity I on a surface is determined by the cosine of θ, the angle of incidence of the light beam to the surface normal, and the intensity I_S of the light source; *max_angle* indicates the angle where the light intensity is just sufficient for growth.

in light intensity ($L(\theta) = I/I_S$). The *generator processing function* is displayed in (3.20).

$$\begin{aligned} L(\theta) &= \cos(\theta) \\ l &= s \cdot L(\theta) \end{aligned} \tag{3.20}$$

In form I_1 no reflection from the environment ($max\_angle = \pi/2$) was used. In form I_2 the same construction was applied, only the parameter *max_angle* was set to π. In both I_1 and I_2 the two coherence conserving rules, *insertion* and *continuity rule*, are applied to ensure that the tangential elements remain connected.

The Biological Objects. In the autotrophic example *Montastrea annularis* (see Fig. 3.9) growth velocities are highest at the tip of the column-shaped colony, and the corallites are secreted superficially. Growth of the corallites is related to the angle θ of the corallite with respect to the light source (Graus and Macintyre 1982). In this case the vertical is the axis of growth and the length l of new longitudinal elements in (3.20) is determined by the light model, the function $L(\theta)$ (3.19). When there is enough reflection from the environment a hemispherical form emerges (comparable with the simulated form Fig. 3.17I_2). In form I_2 *max_angle* in the light model $L(\theta)$ is larger than $\pi/2$. When the light intensity and reflection decrease, the form transforms into Fig. 3.17I_1. The latter form can be used as a simulation of the actual object shown in Fig. 3.9. In Fig. 3.17I_1 *max_angle* is about $\pi/2$. The angle *max_angle* represents

Maximum growth angle

the angle which the longitudinal element of a corallite can make with the vertical and where the light intensity is just sufficient for sustaining the photosynthesis. In the case of this autotrophic example the influence of the physical environment on the growth process is modelled by the light model. For the autotrophic organisms with external secretion and with radiate accretive growth, this function $L(\theta)$ is the basic cause of the radiate accretive structure. Growth of *Montastrea annularis* can be modelled with a subset of the rules discussed so far, where only the coherence conserving rules (*continuity* and *insertion rule*) and the light model $L(\theta)$ are relevant. Longitudinal sections of ecotypes of this coral can be simulated by adapting the *max_angle* in the light intensity function, as shown in Fig. 3.17I_1 and I_2.

3.7.2 A Combination of the Geometric Model and the Concentration Gradient Model

Diffusion process

The Model. The nutrient concentration c in a diffusion process can be described with the Laplace equation. It is possible to determine the concentration c in a field where an object grows and is consuming the nutrient (Meakin 1986). The concentration c is zero on the object and it is assumed that the diffusion process is fast compared to the growth process. The nutrient concentration remains fixed ($c = 1.0$) at the top of the field and is zero at the bottom, when a linear source of nutrient is assumed (Meakin 1986). The concentration field will reach a steady state, where the local concentrations are described by a solution of the Laplace equation (see Sect. 2.4). The growth of an object in a concentration gradient can be simulated in a two-dimensional lattice. The concentrations of nutrient can be approximated for all sites in the lattice with the approximation algorithm in (2.9).

Nutrient concentration distribution

If it is assumed that an object as shown in Fig. 3.17 consumes nutrient from its environment, and if nutrient is supplied by a diffusion process, the same method as applied in the DLA model can be used to determine the nutrient concentration distribution around the growing object (see also Kaandorp 1993a). The algorithm in which the nutrient distribution is determined is described in (3.21). In the algorithm a combination of the geometric model *objects* and lattice representations of the object (*lattice_prev* and *lattice_updated*) is used. The lattice representation of the object is necessary for solving the Laplace equation with a similar algorithm as (2.9). In the examples below, two lattices each with 1000^2 sites are used. The sites in the lattices contain the values of the local nutrient concentration, which varies in the range 0.0..1.0. Initially all sites in *lattice_updated* are set to the maximum concentration 1.0. The sites in the lattice can be in two possible states: "occupied" and "unoccupied". In the actual implementation the state "occupied" is represented by a number larger than 1.0, which is added to the local nutrient concentration in the lattice site. (3.21)

```
det_nutrient_distribution( object, lattice_prev, lattice_updated ){
   step A (erasing previous lattice representation object ):
   for each lattice site with lattice coordinates i, j {
      if (lattice_updated[i][j] ≥ "occupied")
         lattice_updated[i][j] = 0.0; }

   step B (mapping the object on the lattice):
   the geometric model object is mapped on the lattice lattice_updated;
   lattice_updated is copied to lattice_prev;

   step C (solving the laplace equation):
   do {
      completed = TRUE;
      for each lattice site with lattice coordinates i, j {
         prev_value = lattice_prev[i][j];
         if (prev_value ≥ "occupied") updated_value = "occupied";
         else if (j == lattice_size – 1) updated_value = 1.0;
         else if (j == 0) updated_value = 0.0;
         else {
            updated_value = 1/4 · (conc(i+1,j),conc(i–1,j),conc(i,j+1),conc(i,j–1));
            updated_value = (1 – ω) · prev_value + ω · updated_value; }

         if ( (prev_value – updated_value) > tolerance ){
            lattice_update[i][j] = updated_value;
            completed = FALSE; }
         else lattice_update[i][j] = prev_value;
      }
      copy lattice_updated to lattice_prev;
   } while (! completed );

   step D:
   for each tangential element in the state F in object {
      an edge probe is drawn perpendicular to the tangential element;
      an estimation is made of the local nutrient gradient using (3.23);
      the tangential element is associated with this estimation; }
} end det_nutrient_distribution
```

The algorithm is applied after each iteration step and can be divided into four steps. In the first step the representation of the object in the preceding iteration step is erased, by setting the lattice sites in the state "occupied" to zero. In step B the object is mapped on *lattice_updated*. The boundary is mapped onto the lattice by drawing the edges with a modified version of the Bresenham algorithm (Foley et al. 1990). This algorithm is usually applied for drawing line segments, visualized in pixels, on a pixel screen. The contour within the boundary, consisting of sites in the state "occupied", is filled by using a scan-line algorithm (Foley et al. 1990).

Mapping the object onto the lattice

In step C the Laplace equation is solved in a slightly different way as in the approximation algorithm of (2.9). When the last algorithm is used, the process converges very slowly. Step C is computationally the most expensive step. In order to speed up the convergence in step C, systematic over relaxation was used (see Ames 1977; Press et al. 1988). In step C two lattices, *lattice_prev* and *lattice_updated*, are used: *lattice_prev* with the previous states, while *lattice_updated* contains all updated values. The new value *updated_value* of a lattice site depends on the j-coordinate. A linear nutrient source is assumed and all top lattice sites (j-coordinate is lattice_size – 1) are set to the value 1.0. All bottom lattice sites are set to the value zero. Lattice sites which are part of the object and in the state "occupied" are unchanged in the iteration process in step C. A neighbouring site which is in the state "occupied" counts as a zero value in the approximation process by using the function:

Systematic over relaxation

```
conc( i, j ){                                                    (3.22)
   if ( lattice_prev[i][j] ≥ "occupied" ) return( 0.0 );
   else return( lattice_prev[i][j] );
}
```

The new value of a lattice site which is not situated at the object or the bottom or top row is determined by the average value of its 4-connective neighbours and the previous value *prev_value*. The contribution of *prev_value* to *updated_value* is controlled by the relaxation parameter ω, with values: $1 \leq \omega < 2$. In the simulations ω was set to the value 1.9 to attain a fast convergence. The approximation process in step C converges as soon as the difference between *prev_value* and *updated_value*, for all lattice sites, drops below the threshold *tolerance*. In the simulations *tolerance* was set to the value 0.001.

In the final step an estimation was done of the local nutrient concentration gradient. This was done by drawing an edge *probe* with the Bresenham algorithm in the lattice. The edge *probe* is set perpendicular

Local nutrient concentration gradients

to a tangential element in the state F and points into the environment surrounding the object. The values of the sites situated on the edge are used to estimate the gradient. In this estimation an exponent η (3.23) was assumed to describe the relation between the local field and the concentration (Niemeyer et al.1984, Meakin 1986).

$$k(c) = c^{\eta} \tag{3.23}$$

By setting η to a value below 1.0, a steeper overall nutrient gradient is realized, in the simulations η was set to the value 1.0. The nutrient gradient was used in the calculation of the length l of new longitudinal elements in the iteration process (Fig. 3.18) by multiplying the product $f(\alpha) \cdot h(rad\_curv)$ ((3.11) and (3.14)) with the function $k(c)$ (3.23). This combination is shown in the *generator processing function* in (3.24).

$$l = \begin{cases} s \cdot f(\alpha) \cdot h(rad\_curv) \cdot k(c) \text{ for} \\ f(\alpha) \cdot h(rad\_curv) \cdot k(c) > inhibition\_level \\ 0.0 \text{ for } f(\alpha) \cdot h(rad\_curv) \cdot k(c) \leq inhibition\_level \end{cases} \tag{3.24}$$

Visualization of nutrient concentration distribution

A combination of the geometric model and the nutrient concentration gradient obtained by applying the generator processing function from (3.24) is depicted in Fig. 3.17J. More evolved objects generated with this model are shown in Fig. 3.27. In both pictures the basins of equal ranges of nutrient concentration are visualized as alternating black and coloured regions (Mandelbrot and Evertsz 1990, compare Fig. 2.15). The nutrient concentration decreases when the black or coloured basin is situated closer to the object, the decrease in concentration in the coloured basins is shown by a shift in colour from blue to white. The growth layers in the object are visualized with brown colours, the basin in which the object is located, with concentration near zero, is displayed in black. In object Fig. 3.27A the value of max_curv (3.14) was set to $7.2s$, η (3.23) was set to 1.0; in object Fig. 3.27B the parameter max_curv was set to the value $21.7s$, while η was set to 0.5.

Negative substrate-tropism

The Biological Objects. Next to the secretion of new layers of growth proportional to the angle of the axis of growth, and tip-splitting, two more aspects in the growth process of the heterotrophic example *Haliclona oculata* can be observed: the samples in Fig. 3.3 show negative substrate-tropism and suppression of growth of the shielded branches by the canopy of branches.

In the range of sponges shown in Fig. 3.3 it can be observed that there is a clear tendency in the sponges to grow from the substrate. The same effect can be seen in the case when the sponge grows on the ceiling

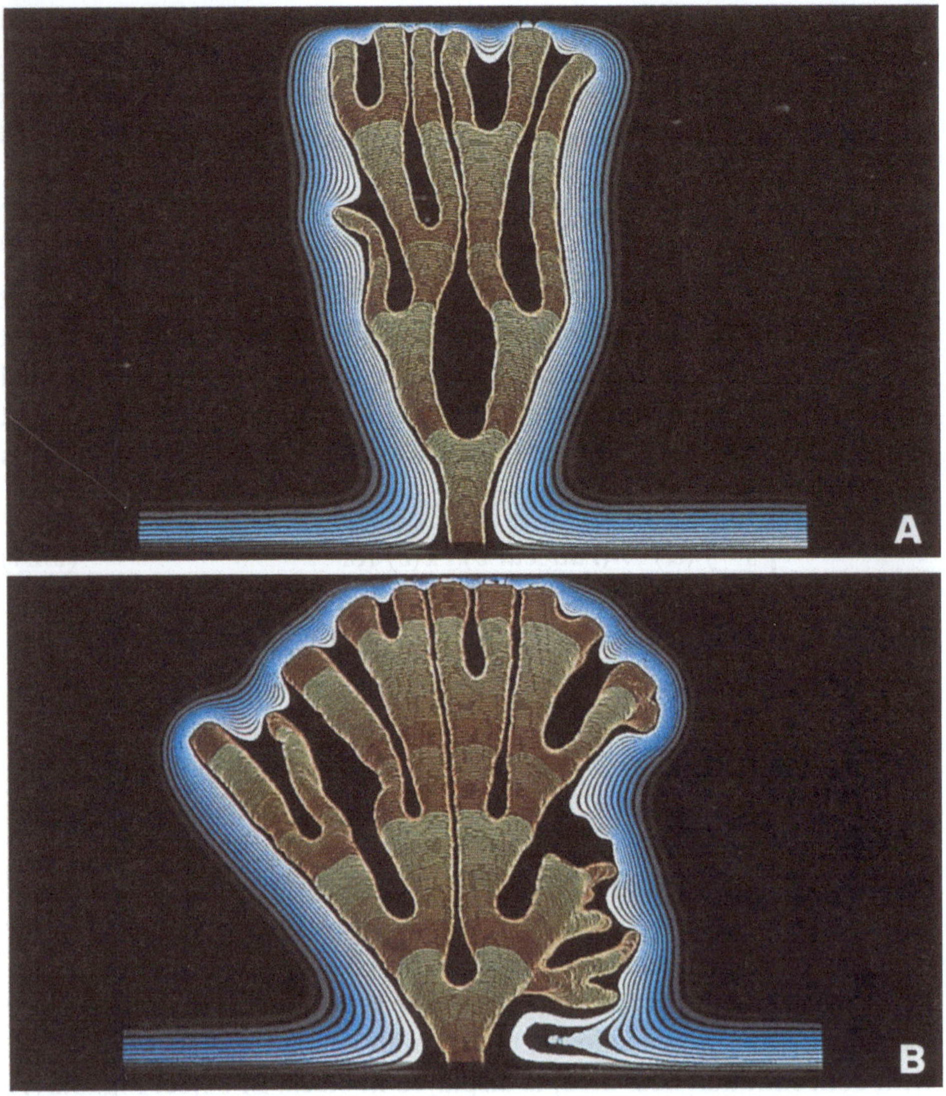

Fig. 3.27. (A) Simulated form generated with a combination of the geometric and the DLA model, using the *generator processing function* in (3.24). The parameter max_curv (3.17) was set to $7.2s$, while the parameter η in (3.23) was set to the value 1.0. The basins of equal ranges of nutrient concentration are visualized as alternating black and coloured regions. The nutrient concentration decreases when the black or coloured basin is situated closer to the object and this decrease is vizualized in the coloured basins as a shift from blue to white. The growth process in the object itself is visualized by using different brown colours for the growth layers. (B) Simulated form generated with a combination of the geometric and the DLA model, the parameter max_curv was set to the value $21.7s$ and η was set to the value 0.5. The same colour setting as in the object in (A) was used.

of a cave. In general the maximal angle between the axis of growth (see Fig. 3.28) and the vertical will be smaller under sheltered conditions when compared to the exposed conditions. Under sheltered conditions this negative substrate-tropism is stronger and can be measured using the maximal angle shown in Fig. 3.28. Negative substrate-tropism can be explained hydrodynamically: assuming a laminar flow, the water movement is zero just at the (fixed) substratum and increases quadratically with the distance from the substratum until the velocity of the laminar flow is reached (e.g. Vogel and Bretz 1971, Vogel 1983). This causes the sponge to grow from the substrate towards the area with the highest flow velocities and the highest supply of suspended material. In sheltered conditions the vertical

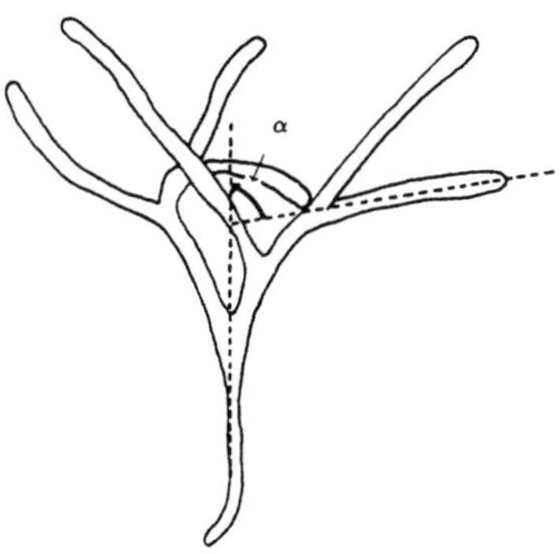

Fig. 3.28. Determination of the maximal angle between an axis of growth of a sponge, or simulated object, and the vertical

gradient in water velocity (together with the food supply and resulting growth velocity) will be steeper than in exposed conditions, resulting in a smaller maximal angle between an axis of growth and the vertical, when compared to the exposed situation.

The aspect of suppression of growth in a shielded branch is experimentally demonstrated in the longitudinal section of Fig. 3.12. This section reveals that growth stops in the right branch as soon as the (marked) left branch overgrows the right branch. Especially on a sheltered site, food supply as well as growth velocity decreases when branches collide. In the samples shown in Fig. 3.3 the lower branches are significantly shorter than the upper branches, since growth is suppressed in these lower branches because of this shielding effect.

In order to model growth which exhibits negative substrate-tropism and suppression of growth in the shielded branches, it is necessary to add a model of the physical environment which describes the distribution of nutrients. In physics the DLA model has been applied to explain fractal growth patterns, as for example found in electrodeposits (Brady and Ball 1984, Sander 1986) and to describe the concentration of particles around the growing objects. In the models shown in Fig. 3.27 this method is used to simulate the distribution of suspended material around a filter-feeding organism like a sponge under sheltered conditions. Under these conditions the suspended material can be considered as randomly moving particles. The particles are only of a larger scale than for example occur in a electrolytic solution. In a DLA model with a linear nutrient source, the negative substrate-tropy can be modelled (see Fig. 2.15). With a combination of this model and the geometric model (3.24) these effects can be simulated. This combination yields the simulated forms of Fig. 3.27, which exhibit negative substrate-tropism and a suppression of growth of the shielded branches by the canopy of branches.

Modelling anti-collision rules in a natural way

With the suppression of the growth of branches which approach each other too closely and deplete the same nutrient source, the anti-collision rules in the iteration process (*non-intersection* and *avoidance rule*) can be partially captured in a more natural way. In Fig. 3.27 it can be seen that the nutrient is locally depleted at sites enclosed by branches; this effect suppresses growth of shielded branches. The diffusion model holds only under sheltered conditions; under exposed conditions food supply will be higher at the parts of the sponge shielded by the canopy of branches from the environment. Under exposed conditions collision of branches may lead more often to anastomosis. This aspect of the growth is not (yet) included in the model.

In Fig. 3.27 a simulation of the growth process with a nutrient distribution under sheltered conditions is depicted. The diffusion model can describe the situation under sheltered conditions. Under exposed conditions laminar and turbulent flows will disturb this pattern. In many marine organisms it can be observed that a flattened growth form emerges, where the flattened plane is perpendicular to the flow direction. A flattened growth form can develop because the growth velocities are maximal when the direction of growth is perpendicular to the flow direction and minimal when both directions are parallel. To model the emergence of these flattened forms it is necessary to extend the model to 3D and to introduce the influence of the flow direction and drift of nutrients on the growth process. To describe this situation accurately it is necessary to use the Navier-Stokes equations instead of the Laplace equation. In Chap. 5 the influence of the direction of flow in a model of radiate accretive growth will be discussed in more detail.

Flattened growth forms and the direction of the flow

Of course a crucial simplification is that the growth process is modelled in 2D. In reality the situation that the nutrient is depleted between the branches (see Fig. 3.27) will occur less frequently because nutrient will be supplied from more directions.

In general, negative substrate-tropism and suppression of growth of the shielded branches by the canopy of branches will occur frequently among branching marine sessile organisms (see also Fig. 3.8). However, for a robust simulation model, a model of the influence of the nutrient distribution around the simulated organisms is an essential part.

3.8 Conclusions and Restrictions of the 2D Model

Fig. 3.12 reveals that growth ceases in the right branch as soon as the (marked) left branch overgrows the right branch. In the model in which the distribution of nutrients around the object was included (Fig. 3.17J) it was assumed that especially on a sheltered site water movement and food supply as well as the growth velocity decreases when branches collide. The last photograph (Fig. 3.12) is experimental evidence for this effect. Under exposed conditions this effect is less critical: because of a relatively higher degree of water movement, food supply will be higher at the parts of the sponge which are shielded by the canopy of branches from the environment. The diffusion model can describe the situation under sheltered conditions, but under exposed conditions laminar and turbulent flows will disturb this pattern. To describe this situation accurately it is necessary to replace the Laplace equation (2.7) by a model which includes the flow direction and drift of nutrients on the growth process.

Collision of branches

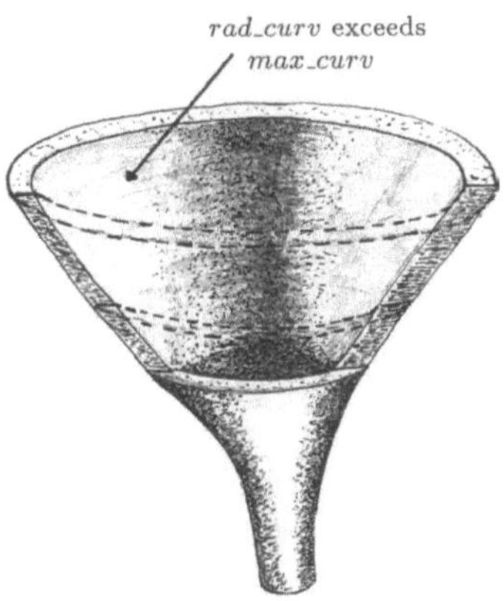

Fig. 3.29. Example of an object in which the function from (3.14) delivers an incorrect estimation of the contact of the indicated parts of the object with the environment

In the *generator-processing functions* (for example (3.24)) in which the influence of the limitations of the transport system are included (3.14) the radius of curvature rad_curv is used to estimate the contact between the tissue and the environment. This method works for the objects generated with the models shown below. In forms as shown in Fig. 3.29 it can be seen that in in some parts of the tissue rad_curv will become very large and exceed the maximum limit max_curv, while these parts are in short distance contact with the environment. For this type of object another type of function is necessary to estimate the contact with the environment.

Deviations from the trend, thin-branching to plate-like forms caused by a difference in degree of exposure to water movement, easily arise as a result of damage of the sponge in the course of time. The more irregular form in Fig. 3.3E is typical for sponges with an age of several years. Probably tissue-material has been removed by abrasion and irregularities have arisen when new material was added. This aspect of the growth process would demand a further extension of the 2D model, for example by introducing a fourth rule in the iteration process that represents removal of material especially from the protruding parts of the object, in order to simulate the process of abrasion.

In Figs. 3.25A, 3.25B, and 3.25C it can be seen that the degree of branching increases when the objects become more plate-like; in the same sequence the branches exhibit a higher degree of colliding. In the model, growth ceases as soon as branches are at the point of intersection. In reality (see Fig. 3.3) the plate-like forms also exhibit a higher degree of branching, but growth continues because branches that intersect can avoid each other in the third dimension. Under natural conditions, especially on a sheltered growth site, water movement and food supply as well as the growth velocity will decrease when branches collide. Under more exposed conditions this last phenomenon is less critical. Sponges from exposed sites exhibit a high degree of anastomosis and a complicated branching system. Due to anastomosis a strong construction is formed, which may withstand strong water movements. The anastomosing surfaces are interconnected and form a new continuous surface on which growth continues.

Anastomosis

A major improvement will be the extension of this model to 3 dimensions. As mentioned in Sect. 3.5, the skeleton of *Haliclona oculata* shows a radial symmetry. Some of the processes (for example the formation of branches, plate-forming at the extremities of branches) can be described with a 2D model because of this symmetry. Other processes, like colliding of branches and anastomosis, can be adequately described only with a 3D model. The growth model is based on the skeleton architecture and the aquiferous system. Both represent the basal patterns in growth forms.

Together with the environmental factors they are the main parameters in causal explanations of growth forms of sponges.

Autotrophy and heterotrophy

It is useful to create a model for radiate accretive growth which unifies autotrophic and heterotrophic organisms. With such a generic model it is possible to model growth of organisms where light as well as the heterotrophic nutrient source are the main environmental parameters which determine the growth form, a situation which occurs among many Scleractinia (Bythell 1988) and some Porifera (Wilkinson et al. 1988). The formation of branches can indicate a significant contribution of the heterotrophic component. For organisms which exhibit a combination of autotrophic and heterotrophic component, the simple light model as shown in (3.20) will not be sufficient in many cases. In the case where branching appears, cast shadows will suppress growth in the lower over-shadowed branches. This can only be modelled adequately with a 3D geometric model and a light model which takes cast shadows into account. A model based on ray-tracing techniques (not as usually applied in computer graphics but a physical illumination model), in which estimations are done for the local light intensities for each growing element, could be a good reflection of the actual environment.

3.9 List of Symbols Used in this Chapter

α	angle between an axis of growth and a tangential element
$f(\alpha)$	function describing the deposition of a new layer of tangential and longitudinal elements
l	length of a longitudinal element
s	length of tangential element
M_{ij}	matrix operator i, using homogeneous coordinates, in iteration step j
V_i	a vertex with coordinates V_{xi} and V_{yi}
$T(DX, DY)$	translation over the vector $[DX, DY]$
F, SF, NF	an edge can be respectively in the state: "fertile", "non-active" or "not-fertile"

$b0, b1$	vertices of new tangential element constructed from the tangential edge(V_i, V_{i+1})
pb	midpoint of the edge($b0$, $b1$)
$a0, a1$	vertices of an edge which is the left adjacent tangential edge of edge($b0$, $b1$)
pa	midpoint of the edge($a0$, $a1$)
$c0, c1$	vertices of an edge which is the right adjacent tangential edge of edge($b0$, $b1$)
pc	midpoint of the edge($c0$, $c1$)
$n0, n1$	vertices of a new tangential edge which is inserted between edge($b0$, $b1$) and edge($c0$, $c1$)
$inhibition\_level$	threshold below which l becomes zero
w	widening factor in $f(\alpha)$
$g(lowest\_value)$	function returning random values between the limits $lowest\_value$ and 1.0
pv	midpoint of a tangential edge which is not the same edge as ($a0$, $a1$),($b0$, $b1$) or ($c0$, $c1$)
rad_curv	radius of curvature formed by a set of 3 points situated on neighbouring tangential elements
$h(rad\_curv)$	function which returns a normalized version of the radius of curvature
min_curv	minimum value radius of curvature (constant value), expressed in units s
max_curv	maximum value radius of curvature, expressed in units s
θ	angle of incidence of the light beam and surface normal
max_angle	maximum angle which a longitudinal element can make with the vertical
$L(\theta)$	function which models the influence of the light intensity on the growth process
η	exponent describing the relation between the local field and the concentration c
c	local nutrient concentration
$k(c)$	function which represents the influence of the nutrient distribution on the growth process

4 A Comparison of Forms

In the first section of this chapter the simulated forms and the actual growth forms are compared to each other. A quantitative comparison of these forms is an essential step in testing the biological validity of simulation models. When both virtual and actual forms can be quantified, is is also possible to determine a relation between the model parameters and the observed forms.

In the second section the effect of changing an environmental parameter, in an experiment, on the growth form is tested. The effect on the growth form is quantified with the methods described in the first section. These experiments are used to test the predictive value of the simulation model.

4.1 A Comparison of a Range of Forms

In this section the ranges of growth forms found along a gradient of an environmental parameter, as well as the ranges of simulated forms, are compared to each other. The range of growth forms of *Haliclona oculata* (Fig. 3.3) is used as an example, the same methods can also be applied for ranges of growth forms of other organisms (for example Fig. 3.8) as well as for a simulated range of forms in which one or more parameters are gradually changing (see Figs. 3.25A, 3.25B, and 3.25C). For this comparison two different approaches are used. In the first approach the self-similar aspects of the forms are compared. The choice of a suitable self-similar aspect, for example the self-similarity ratio, depends on the types of object which are to be compared. Such a comparison is only useful when a class of related objects is considered. A class of related objects can be generated by using one type of generator. The generators discussed in Sect. 2.6, on iterative geometric constructions, are all based on a linear transformation, which can be written as a combination of scaling, rotation, and translation. If one parameter is gradually changed in this linear transformation, one obtains a set of generators which result in a class of related objects. In the example of the series of ramifying objects (Fig. 2.35) such a related class of objects is shown. The self-similar aspect can be expressed as the self-similarity ratio t/b in Fig. 2.33. In this example the scaling factor in the linear transformation embodied by the generator is gradually changed. The determination of self-similar aspects in a given image follows basically the same approach as discussed in Sect. 2.5 on Iterated Function Systems, where these aspects are described in a set of mappings. In the second approach fractal dimensions of the forms are determined. In a comparison of forms this fractal dimension is again only useful when a class of related objects is compared. Many totally different objects may be characterized by the same fractal dimension.

Fractal dimensions

In order to enable a comparison between the actual growth forms and the simulated 2D forms, projections (photographs) perpendicular to branching plane of the organisms were made. The range of forms of a branching organism originating from a gradient in environmental parameters, as well as a range of simulated objects, can be compared by measuring the diameters of circles just fitting in the branches. In Fig. 4.1 the determination of these circles in the contours of a photograph of a growth form of *Haliclona oculata* is shown. Diameters of two circles were measured: the diameter da of the largest circle (a) which fits in the branch just before ramification and the diameter db of the largest circle (b) just after ramification. Additionally the distance rb was measured between the centres of two successive circles b and a. This distance expresses the length of a branch.

Measuring the diameters da and db

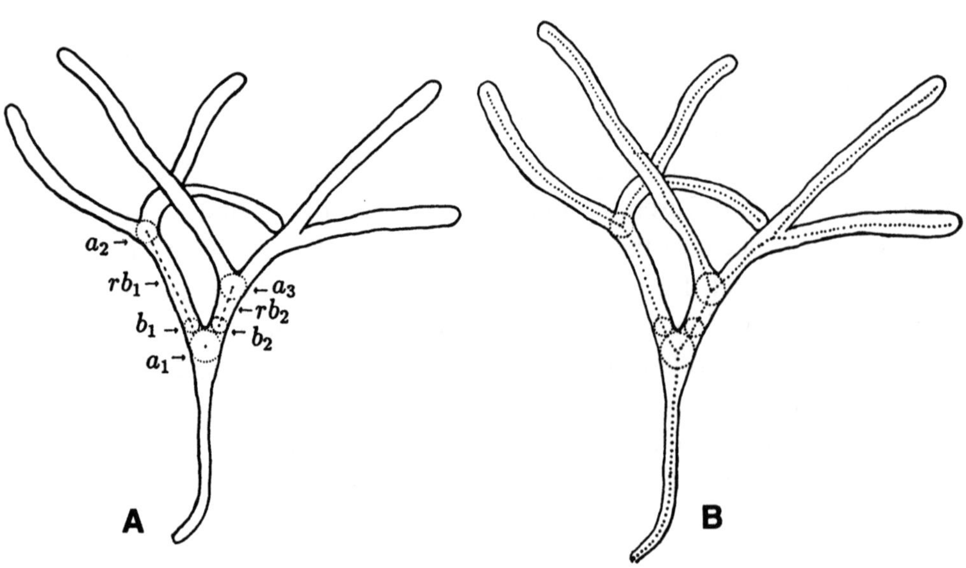

Fig. 4.1. Diagram of the contours of a sponge, with the locations of the measured circles. A: circles a are the largest circles which fits in the branch just before ramification, b are the largest circles just after ramification, and rb is the distance between the centres of two successive circles a and b. B: the skeleton of the object is constructed by connecting the centres of a series of maximum discs.

Maximum discs

The circles a and b correspond to the "maximum disc" from mathematical morphology (see Rosenfeld and Kak 1976). The maximum disc is defined as a disc which fits within the contours of the object and which is not enclosed by another disc fitting within the contours. The (morphological) skeleton of an object, as shown in Fig. 4.1B, can be constructed by connecting the centres of a series of maximum discs. The circle a can be defined as a maximum disc whose centre is positioned at a branching point of the skeleton.

4.1.1 A Comparison of a Range of Actual Forms and the Virtual Objects

Estimating the exposure to water movement

In the next two sections growth forms of *Haliclona oculata* from localities with a different exposure to water movement are compared to each other and to the virtual forms. Details about the measurements and locations can be found in Kaandorp 1991b. The differences in exposure to water movements in these locations is estimated by relating the erosion of gypsum blocks to the rate of exposure. The erosion value is expressed as the weight loss of the gypsum blocks (g/hour) during a lunar day (24.45 hours). The erosion values for the different locations, discussed in the next two sections, is given as an indication of the differences in exposure to water movement (see De Kluijver 1989).

The measurements da, db, and rb

In Fig 4.2 the diameters da and db and the distance rb, measured in the samples (see Fig. 3.3) from an exposed site (mean erosion value 0.09 g/h) and a sheltered site (mean erosion value 0.06 g/h) and two simulated forms (the thin-branching model in Fig. 3.25A and the plate-like model in Fig. 3.25B) are shown in frequency diagrams. In order to compare both results with each other all measurements on the models are multiplied with the factor $\overline{da}_{\text{sample\_sheltered\_site}}/\overline{da}_{\text{thin-branching\_model\_A}}$. The mean values, standard deviations, and number of observations are summarized in the Table 4.1.

Table 4.1. Mean values and standard deviations of the parameters da, db, and rb (in cm) for the samples and the models (Fig. 3.25A, B)

	$\overline{da}$	s_{da}	$\overline{db}$	s_{db}	$\overline{rb}$	s_{rb}	n
thin branching model A	0.26	0.00	0.13	0.00	1.28	0.59	6
plate-like model B	0.31	0.02	0.19	0.04	0.59	0.33	16
samples sheltered site	0.26	0.05	0.15	0.04	1.97	1.45	102
samples exposed site	0.36	0.08	0.20	0.06	1.32	0.96	131

The values of $\overline{da}$ and $\overline{db}$ are larger for the exposed site when compared with the sheltered site, but $\overline{rb}$ is relatively smaller for the exposed site. The same trend is observed when the plate-like model is compared with the thin-branching model.

The distributions of the measurements of the samples are compared by applying the two-sample rank test (non-parametric). In all cases, the hypothesis that the distribution of the samples from both sites is the same is tested against the alternative that they differ by a translation. In all cases a significance level of 5 % was used. The results of the two-sample rank test are listed in Table 4.2.

Table 4.2. Results of the two-sample rank test, carried out for the three types of measurements done for the samples from the sheltered and exposed site

	result of the test
da	distribution of the samples of the sheltered site is situated left of the exposed one
db	distribution of the samples of the sheltered site is situated left of the exposed one
rb	distribution of the samples of the sheltered site is situated right of the exposed one

Distributions of da and db

It can be seen that the distributions of *da* and *db* of the samples collected at the exposed locality ($\overline{da} = 0.36$ cm) are positioned to the right of those of the samples collected at the sheltered locality ($\overline{da} = 0.26$ cm). The distribution of *rb* for the samples from the exposed locality is positioned left of the one for the samples from the sheltered locality.

A thin-branching growth form (for example Fig. 3.3A) is characterized by a relatively low diameter of the largest circle *a* which fits within the contours, just before branching, together with a low value of the diameter of the largest circle *b* which fits in the contours after branching. This form is characterized by a low degree of branching, which is reflected by a relatively high value for *rb*, the distance between the circles *a* and *b*. The plate-like form (for example Fig. 3.3D) is characterized by a relatively high $\overline{da}$ and $\overline{db}$ and a high degree of branching resulting in a low $\overline{rb}$ (see Tables 4.1 and 4.2). Although there is some overlap in the measured *da* (see Fig. 4.2), plate-like growth forms are more common at the exposed site, while the thin-branching form is typical for a sheltered site.

Fig. 4.2. Frequency diagrams of the diameters *da*, *db*, and the distance *rb*. The diagrams at left side of the figure are the measurements on the samples, at the right those of the models (Figs. 3.25A and 3.25B).

Except for a relatively low $\overline{da}$ and $\overline{db}$ the samples from the sheltered site as well as the thin-branching model (Fig. 3.25A) are characterized

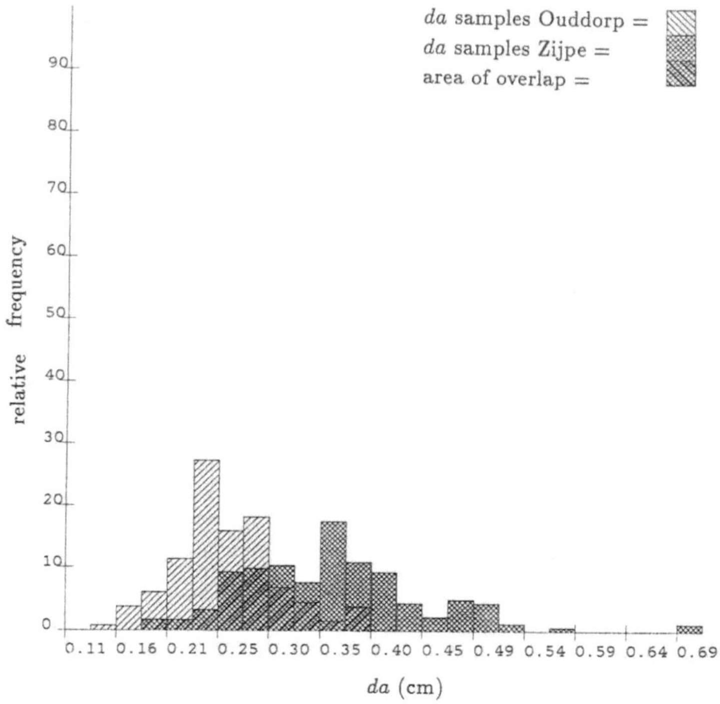
da samples Ouddorp =
da samples Zijpe =
area of overlap =
relative frequency
0 10 20 30 40 50 60 70 80 90
0.11 0.16 0.21 0.25 0.30 0.35 0.40 0.45 0.49 0.54 0.59 0.64 0.69
da (cm)

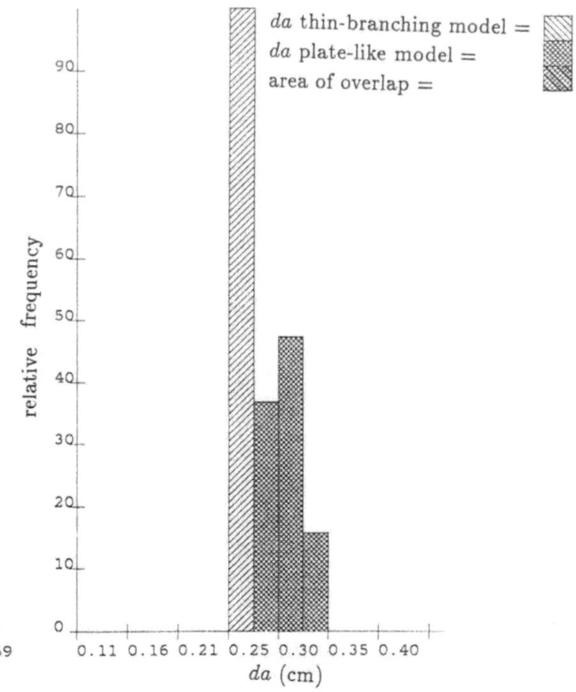
da thin-branching model =
da plate-like model =
area of overlap =
relative frequency
0 10 20 30 40 50 60 70 80 90
0.11 0.16 0.21 0.25 0.30 0.35 0.40
da (cm)

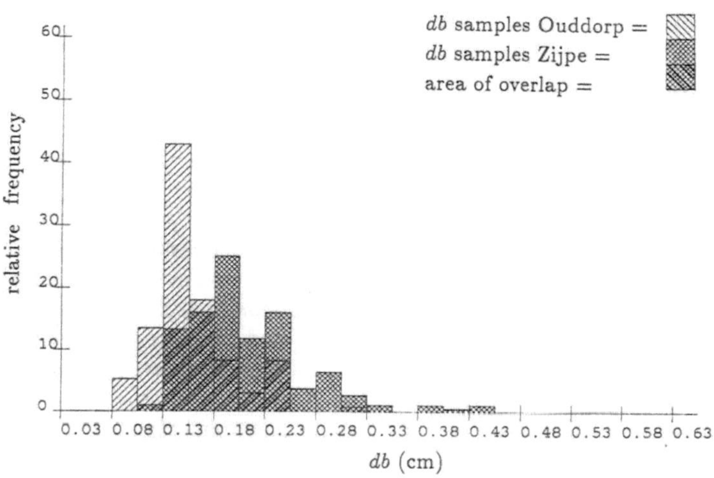
db samples Ouddorp =
db samples Zijpe =
area of overlap =
relative frequency
0 10 20 30 40 50 60
0.03 0.08 0.13 0.18 0.23 0.28 0.33 0.38 0.43 0.48 0.53 0.58 0.63
db (cm)

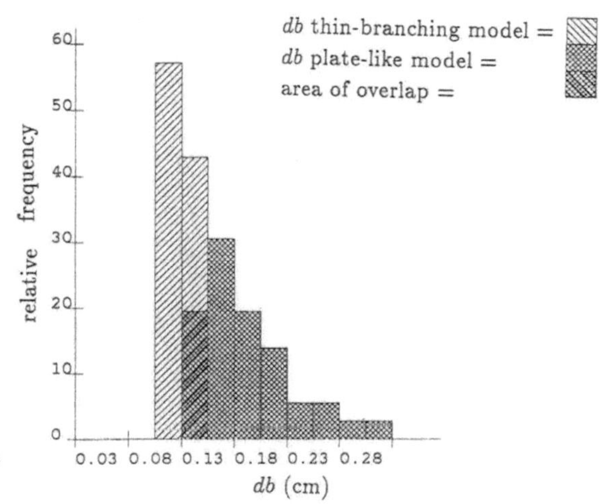
db thin-branching model =
db plate-like model =
area of overlap =
relative frequency
0 10 20 30 40 50 60
0.03 0.08 0.13 0.18 0.23 0.28
db (cm)

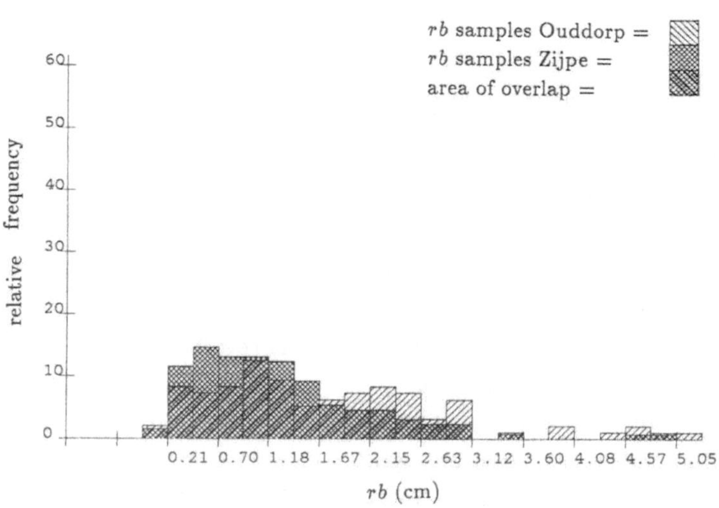
rb samples Ouddorp =
rb samples Zijpe =
area of overlap =
relative frequency
0 10 20 30 40 50 60
0.21 0.70 1.18 1.67 2.15 2.63 3.12 3.60 4.08 4.57 5.05
rb (cm)

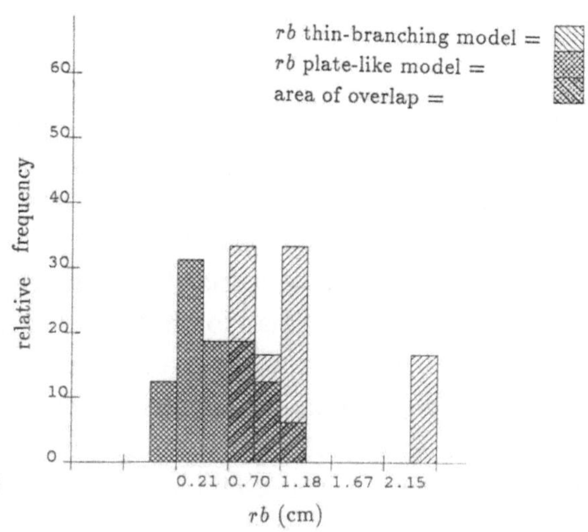
rb thin-branching model =
rb plate-like model =
area of overlap =
relative frequency
0 10 20 30 40 50 60
0.21 0.70 1.18 1.67 2.15
rb (cm)

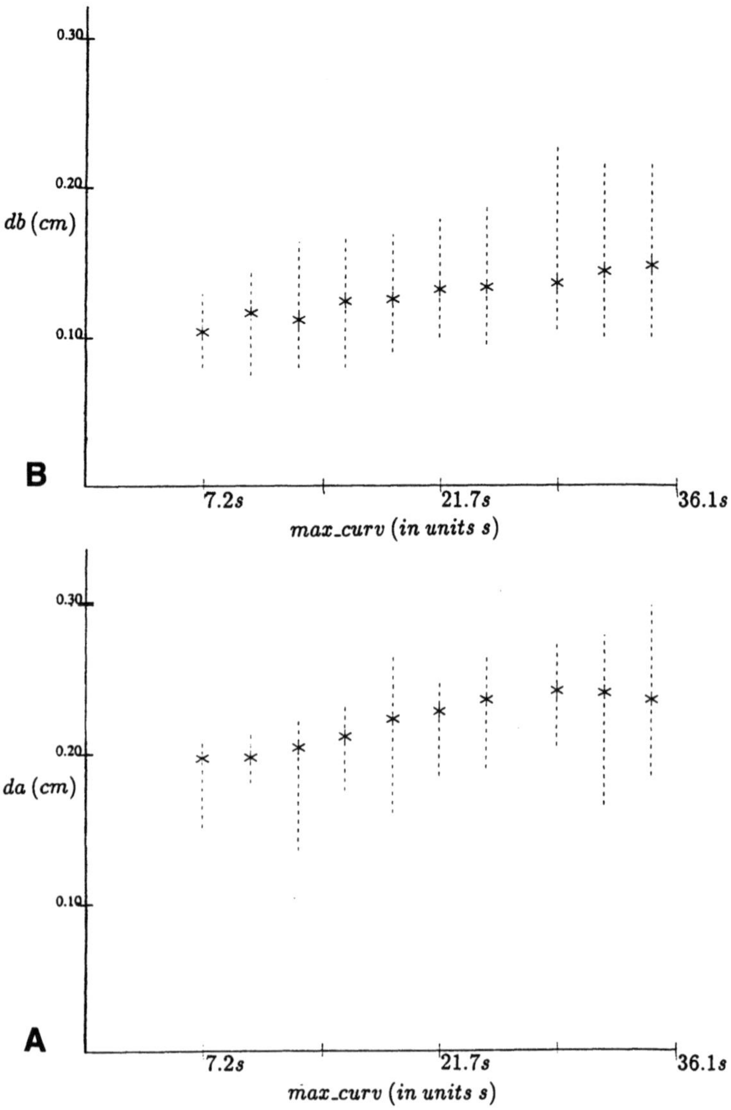

Fig. 4.3. Results of a series of simulations in which the model parameter *max_curv* was varied and the resulting $\overline{da}$ (A) and $\overline{db}$ (B) in the simulated objects was measured. The minimal and maximal values of *da* and *db* are indicated with dotted lines.

by a low degree of branching, with a high $\overline{rb}$ (see Table 4.1). Table 4.1 shows that these values for the thin-branching model and the plate-like model are less extreme than the corresponding values for the samples from the sheltered and exposed locality. From the frequency diagrams it appears that the area of overlap for all three values is larger for the samples collected at both sites than for the thin-branching and the plate-like model.

Degree of plate formation

As said in the preceding section the degree of plate formation can be controlled with the parameters *lowest_value* (3.12) and *max_curv* in the model. In object Fig. 3.25C it is observed that the degree of plate-formation can be increased by using a smaller value for *lowest_value* and larger one for *max_curv*. The degree of plate formation is increased

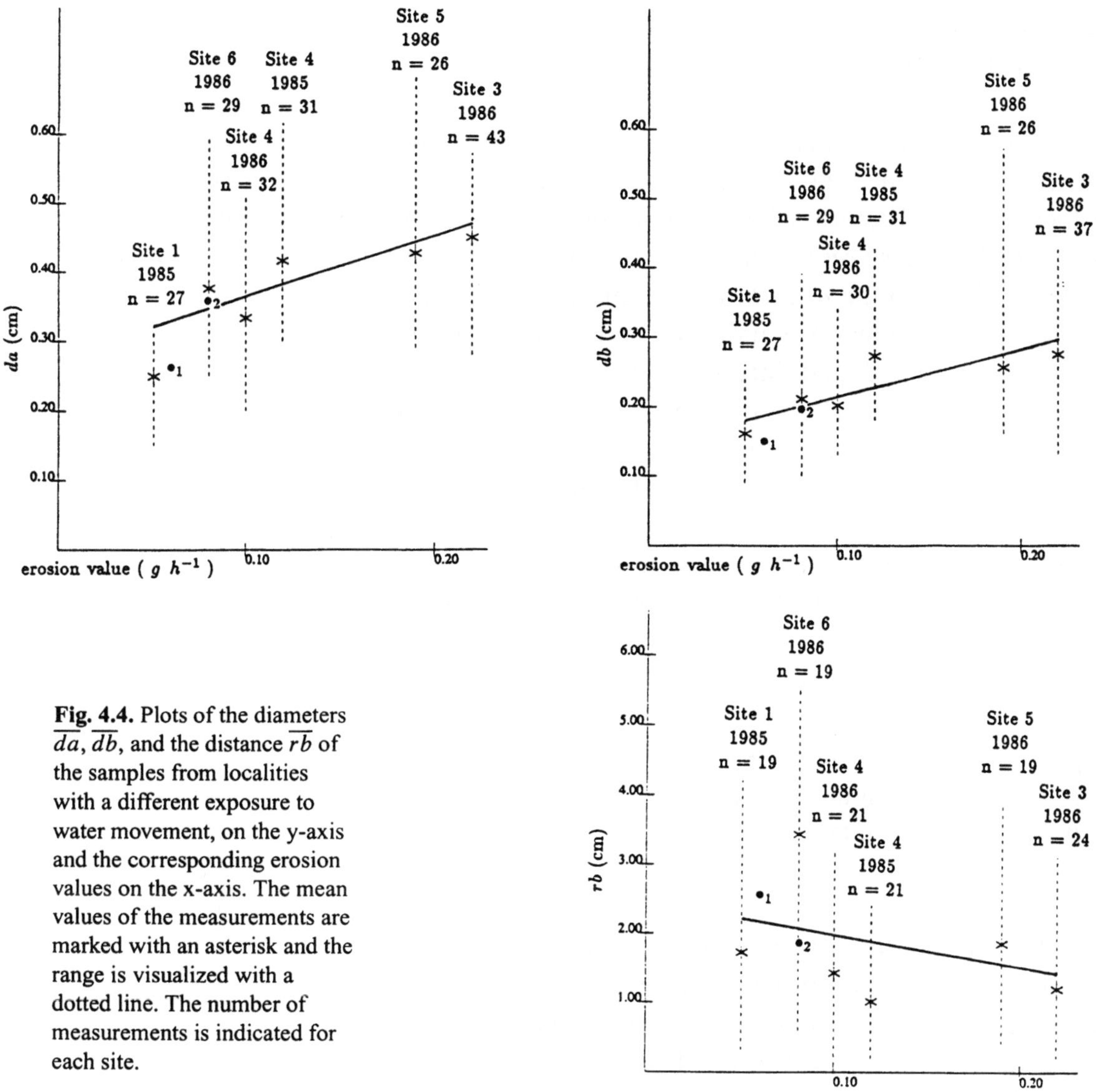

Fig. 4.4. Plots of the diameters $\overline{da}$, $\overline{db}$, and the distance $\overline{rb}$ of the samples from localities with a different exposure to water movement, on the y-axis and the corresponding erosion values on the x-axis. The mean values of the measurements are marked with an asterisk and the range is visualized with a dotted line. The number of measurements is indicated for each site.

Relation between da, db, rb and model parmeter max_curv

resulting in a model with a still higher $\overline{da}$, lower $\overline{rb}$ and higher $\overline{db}$. In Figs. 4.3A and 4.3B the $\overline{da}$ and $\overline{db}$ in simulated objects were determined in a series of experiments in which the model parameters max_curv and $lowest\_value$ were varied simultaneously. In this series of simulations a linear relation, $lowest\_value = 1.5 \cdot max\_curv + 0.25$, was assumed between the two parameters. In the figures it can be seen that both $\overline{da}$ and $\overline{db}$ increase for larger values of max_curv. Both figures are based on 30 measurements for each simulated object, and the da and db values were multiplied with the factor $\overline{da}_{\mathrm{sample\_sheltered\_site}}/\overline{da}_{\mathrm{thin-branching\_model\_A}}$.

4.1.2 A Comparison of the Growth Forms of *Haliclona oculata* Collected in Different Localities

Relation exposure to water movement and da, db, and rb

In Fig. 4.4 the mean values of the measurements of the growth forms collected at sites with a different exposure to water movement (details about these sites located in the Eastern Scheldt in the Netherlands can be found in Kaandorp 1991b) and the regression lines are plotted, assuming that the variance is restricted to the da, db, and rb measurements. For each site the number of measurements is indicated and the range of the measurements is visualized as a dotted line. The regression coefficients were tested to a 5% significance level; the hypothesis that the regression coefficient $\beta = 0$ was tested against the alternative $\beta > 0$ (for da and db), and $\beta < 0$ (for rb). The results are shown in Table 4.3.

Table 4.3. Results of the regression coefficient test, carried out for the three types of measurements done for the samples collected at sites with different exposure to water movement (see Fig. 4.4)

	result of the test
da	hypothesis $\beta = 0$ is rejected in favour of $\beta > 0$
db	hypothesis $\beta = 0$ is rejected in favour of $\beta > 0$
rb	hypothesis $\beta = 0$ is rejected in favour of $\beta < 0$

It can be seen that da and db increase with the erosion value, while rb decreases. The measurements from the exposed and sheltered sites, discussed in the preceding subsection, and their corresponding erosion values are visualized as dots in Fig. 4.4. It can be concluded that plate formation and branching increases with water movement.

4.1.3 Determination of the Fractal Dimensions in a Range of Forms

Contour photographs

In order to estimate the fractal dimensions of *Haliclona oculata*, projections (photographs) were made of the sponges. For this purpose a high contrast film was used, to obtain sharp contours of the sponges. The contours were digitized; two examples are shown in Figs. 4.5A (sheltered location) and 4.5B (exposed location). The contours were digitized by hand-tracing the contours of the photographs. The fractal dimensions of both contours were estimated with the coastline method (Mandelbrot 1983), in which the boundary of the contour is covered with an equal-sided polygon with side

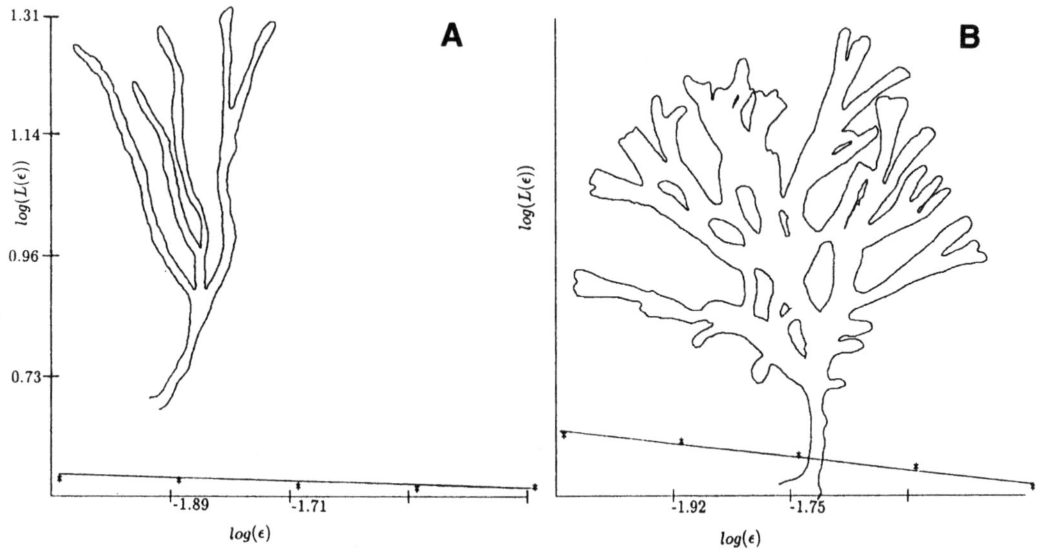

Fig. 4.5. Plots of the total length $L(\epsilon)$ of an equal-sided polygon with side length ϵ on the y-axis and the corresponding ϵ values on the x-axis. The total lengths were measured for the digitized contour shown in the same picture. The exhibited contour in (A) is of a sponge originating from a sheltered site (comparable with Fig. 3.3A). In (B) the same estimation was done for a sponge from an exposed site (Fig. 3.3C).

length ϵ. In Fig. 4.5 these estimations are visualized for both contours. The relation between the total length of the equal-sided polygon $L(\epsilon)$ and the fractal dimension D is given in (2.3). The estimations were done for 6 samples from the sheltered location (mean erosion value 0.06 g/h) and an exposed location (mean erosion value 0.09 g/h) The same estimations were carried out for the simulated objects. In Table 4.4 the estimations of the fractal dimensions of contours of sponges from both sites and of the models in Figs. 3.25A and 3.25B are shown.

Table 4.4. Fractal dimensions, estimated for contours of sponges from a sheltered site, an exposed site, and the thin-branching, plate-like model (Figs. 3.25A and 3.25B)

sheltered site	exposed site	thin-branching model	plate-like model
1.04	1.16	1.26	1.67
1.06	1.14		
1.04	1.19		
1.06	1.08		
1.07	1.20		
1.05	1.11		

The more irregular contours of the sponges from the more exposed site are characterized by a higher fractal dimension ($\overline{D} = 1.15$) than the thin-branching forms form the sheltered site ($\overline{D} = 1.05$). The same tendency can be observed for the plate-like model (Fig. 3.25A, $D = 1.67$)

and the thin-branching model (Fig. 3.25B, $D = 1.26$). In Fig. 4.6 the relation between D and the model parameter max_curv is depicted. It can be seen that that D increases with max_curv. In this experiment the same relation between $lowest\_value$ and max_curv was assumed as used in Sect. 4.1.1. The fractal dimensions are systematically lower for the actual sponges. One possible explanation is that growth in the simulated forms is artificially halted as soon as branches intersect. This limitation of the 2D model increases the irregularity and as a consequence the fractal dimension in the models.

Relation between D and the model parameter max_curv

4.2 An Experimental Verification of the Model

An important aim of constructing simulation models of growth forms is to obtain a better understanding of the way in which these growth forms emerge. In the construction of such models the various aspects of the growth are described in formal rules, leading to a better insight of growth forms, and the parameters in these rules which are responsible for certain aspects in the growth forms can be identified. The identification of these parameters can be used as a base for experiments: if the assumptions made in the models are right it should be possible to verify them by experiments. It is necessary that a growth model not only generates an object that resembles the actual growth form, but that all rules applied in the growth model have a biological significance. Different stages in the growth process can be simulated. An important feature of a correct model is that the effects of changes in the environment on the growth forms can be predicted.

Transplantation experiments

Transplantation experiments form an important method to test the validity of simulation models. In these experiments the normal growth process is interrupted and some of the environmental parameters are modified. A simulation model which takes these parameters into account should be able to predict the changes in growth forms and the final resulting growth form. The growth forms of many marine sessile organisms are strongly influenced by the environmental conditions and can therefore be used for bio-monitoring. This is an important application of simulation models with a well-tested biological relevance. With the actual growth forms and a simulation model, periods of environmental stress, which may lead to interruptions in the growth process, can be retraced in the growth form. In addition to the influence of the exposure to water movement, these interruptions may be caused by the occurrence of pollutants, silt, or other changes in the environment.

Bio-monitoring

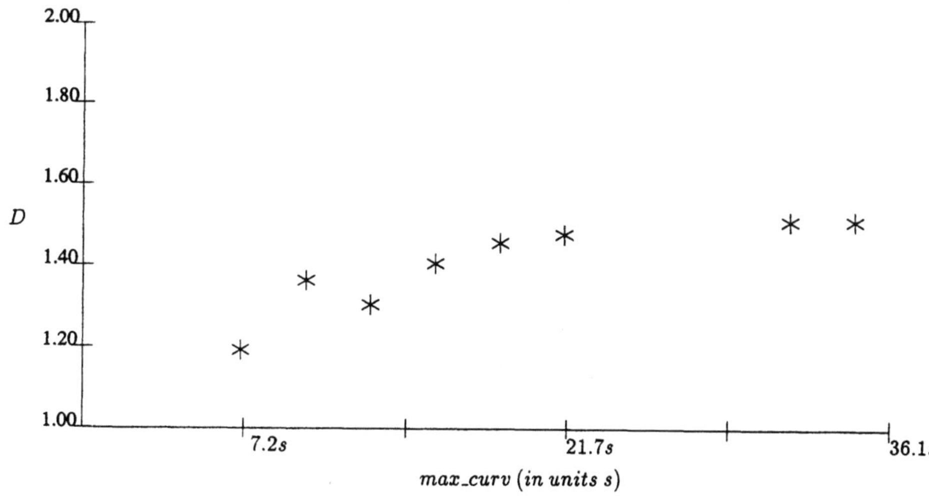

Fig. 4.6. Results of a series of simulations in which the model parameter *max_curv* was varied and the resulting fractal dimension D in the simulated objects was measured

A study in which transplantation experiments were used to obtain insight in the range of growth forms of two closely related species of the hydrocoral *Millepora* was carried out by De Weerdt (1981). Examples of other studies in which transplantation experiments were used for causal explanations of different growth forms of marine organisms, under different environmental conditions, are Graus and Macintyre (1982) on the coral species *Montastrea annularis* (Scleractinia) and Vethaak et al. (1982) on two related species of the sponge *Halichondria.*

4.2.1 The Simulation Experiments

In one set of simulation experiments only two parameters were varied: *max_curv* (3.14) and *lowest_value* (3.12) in the *generator processing function* of (3.17) (the model displayed in Fig. 3.17H). The sheltered conditions were simulated by selecting a low value for *max_curv* and the value 1.0 for *lowest_value*. The exposed conditions were simulated by selecting a higher value for *max_curv* and 0.8 for *lowest_value.*

In Fig. 4.7A and B the forms generated in the simulation experiment are shown. In form A the parameters *max_curv* and *lowest_value* (Fig. 3.17H) were first set respectively to a low value (7.2s) and 1.0 (no disturbance). Without interrupting the iteration process this would result in the thin-branching form shown in Fig. 4.7C. In form A this process was interrupted and disturbed by setting the parameter *lowest_value* to 0.8, and simultaneously *max_curv* was increased (from 7.2s to 21.7s). In form B the reverse experiment was carried out: initially *lowest_value* was set to 0.8 and a high value for *max_curv* was used, and during the

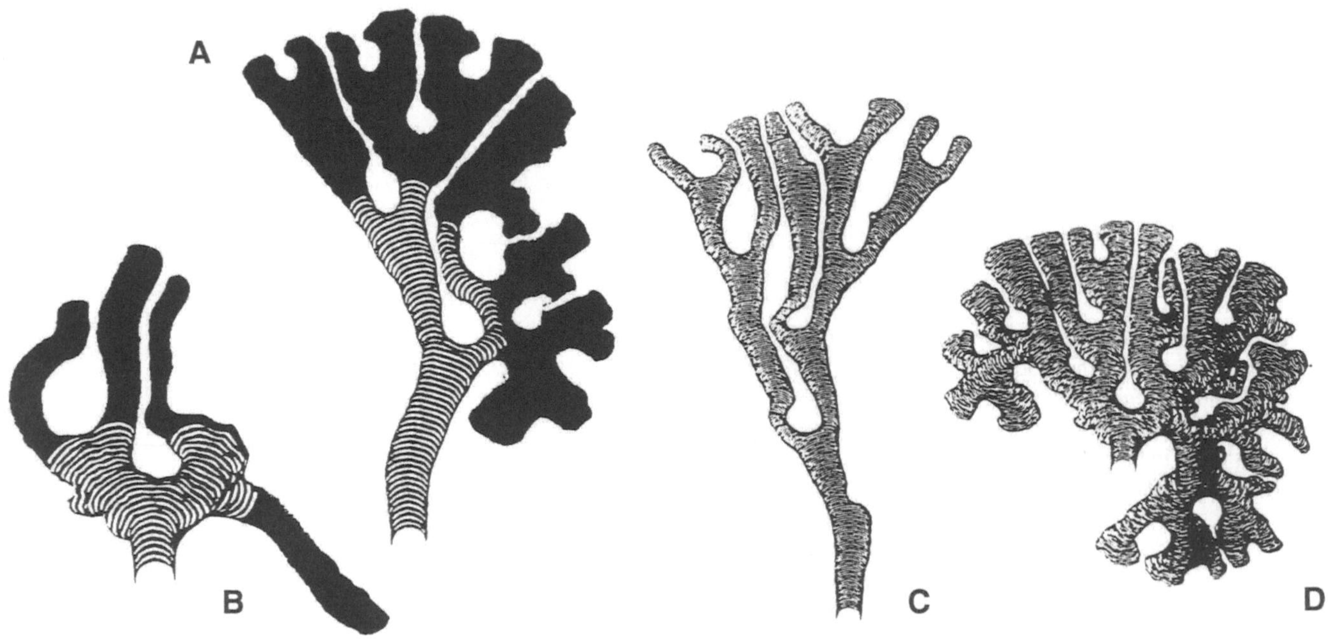

Fig. 4.7. Simulated forms, using the model shown in Fig. 3.17H. In form A the parameters *lowest_value* and *max_curv* were set initially to 1.0 and a relatively low value, then the iteration process was interrupted and the parameters were set to 0.8 and a high value. In form B the reverse experiment was done. Forms C, D result from the iteration process without interruption (parameter settings are the initial ones for A and B, respectively).

generation the parameters were set to 1.0 (disturbance is stopped) and a lower value for *max_curv*. Without this last interruption form D emerges.

In a second set of simulation experiments the effect of the concentration gradient on the form of the object was tested by rotating the object in the iteration process. The *generator processing function* from (3.24) was used (the model shown in Fig. 3.17J). The object was positioned horizontally after 50 iteration steps. In this object the parameter *max_curv* (3.14) was set to 21.7s and the parameter η (3.23) was set to 0.5. The negative substrate-tropism of the model is demonstrated in Fig. 4.8 (see Sect. 3.7.2), where the object grows towards the nutrient source. In the picture the nutrient concentration decreases when the black or coloured basin is situated closer to the object, the decrease in concentration is visualized in the coloured basins as a shift from blue to white, and the object itself is displayed in black.

4.2.2 The Transplantation Experiments

The experiments described in the previous section – transplanting a thin-branching object from a simulated sheltered environment to an exposed environment, the reverse experiment with a plate-like object, and the rotation of an object during the growth process – were also carried out in the field. Details of these transplantation experiments can be found in Kaandorp and de Kluijver (1992).

Fig. 4.8. Form generated with a combination of the geometric and the DLA model using the model in Fig. 3.17J. The object was positioned horizontally after the 50th iteration step. The nutrient depletion around the object the object is visualized by the colour shift from blue to white in the coloured basins, the object itself is displayed in black.

During two periods, sponges were collected from a sheltered site 1 (Lake Grevelingen) and two exposed sites 2 and 3 (both in the Eastern Scheldt, see Fig. 4.9) in the Netherlands. Sites 1 and 2 have rich *Haliclona oculata* populations. Site 3, although poor in *Haliclona oculata* individuals, was chosen because it is an example of a site with a high exposure to water movement. For the experiments small individual sponges were used, in order to avoid complicated growth forms in which the effect of the transplantation can less easily be interpreted.

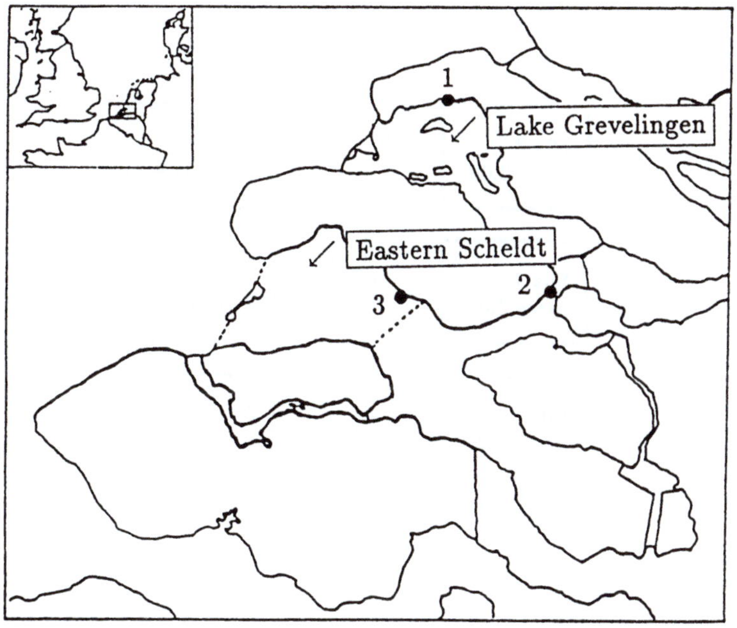

Fig. 4.9. Map of the study area in the Netherlands with the sampled localities (1 is a sheltered location in Lake Grevelingen; 2, 3 are two exposed sites in the Eastern Scheldt)

At all sites some additional sponges were collected before and after the experiments. This material was used for comparing the effects of the experiments on the growth forms as explained below.

In the second experiment an improvement was introduced. In this experiment the sponges were photographed before the transplantation. For exact measurements some branches were marked with minute stainless needles. The needles were stuck into the surface of the sponges, the ends of the needles corresponding with the surface of the sponge before the experiment. After being marked the sponges were fixed to long nails with insulated electricity wire (see Fig. 4.10). Each sponge was supported by a small fissure in the head of the nail in order to ensure that the sponge remained erect. The nails with the sponges were stuck into the substrate at the test sites. After a period of about three months the sponges were recollected and photographed anew, and the position of the needles was detected by using x-ray photographs.

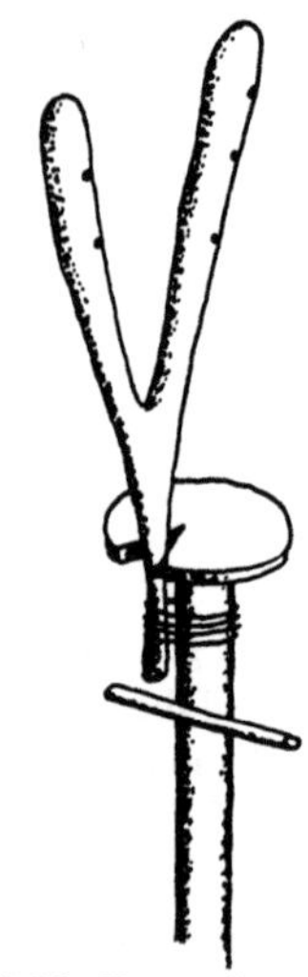

Fig. 4.10. Transplantation experiment with a sponge fixed to a nail

At each experimental site some sponges were returned to the original growth site (the control experiment) and some were transplanted to a site with a different degree of exposure to water movement. The experiment is summarized in Fig. 4.11. In this diagram the number of transplants and the degrees of exposure to water movement are indicated. The erosion values shown in Fig. 4.11 are taken from the literature (see De Kluijver 1989) and are used as an indication of the degree of exposure to water movement on the test sites. The erosion values were measured again during the transplantation experiments on the three test sites. Next to exposure measurements the sedimentation load on the three test sites was determined. Sedimentation is a limiting environmental parameter in the growth process of a sponge; excessive sedimentation (especially fine sediment) may lead to occlusion of the inhalant pores of the aquiferous system (see Brien et al. 1973). For this reason also measurements of the daily sed-

Erosion values

Sedimentation load

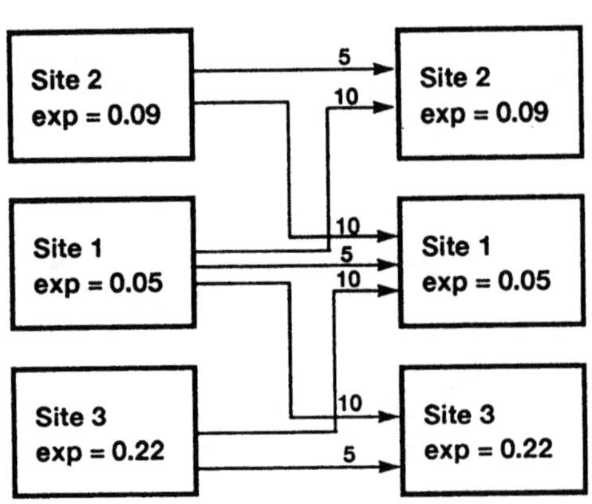

Fig. 4.11. Diagram summarizing the transplantation experiment. The degree of exposure to water movement is indicated for each site (see Fig. 4.9) by the corresponding erosion values (*exp*). The number of experimental individuals is indicated above the arrows.

imentation load were made during the experiments on the three test sites. The sedimentation was measured by using sediment traps and by taking samples of the upper 1 cm layer of the bottom sediment. In the bottom samples a division was in made into different fractions, measured in mm and determined by using 7 graded sieves (2.8–0.05 mm). Details of the sedimentation measurements can be found in De Kluijver and Leewis (in prep.).

Survival rate of the transplants

The survival rate of the transplants is shown in Tables 4.5 and 4.6. The low survival rate of the transplants at site 1 in Table 4.5 is probably caused by human activities. In both tables a relatively low survival rate of the transplants at site 3 is apparent. The erosion values, measured in the period 1989–1990, are listed in Table 4.7. The corresponding sedimentation loads and dominant sediment fractions in the period 1989–1990 are shown in Table 4.8.

Table 4.5. Survival of the transplants in the first experiment (period 11 March 1989 until 30 April 1989)

source	target	number of transplants	number of survivors
site 1	site 1	5	1
site 1	site 2	10	8
site 1	site 3	10	7
site 2	site 2	5	4
site 2	site 1	10	9
site 3	site 3	4	4
site 3	site 1	10	5

Table 4.6. Survival of the transplants in the second experiment (period 30 November 1989 until 17 March 1990)

source	target	number of transplants	number of survivors
site 1	site 1	5	5
site 1	site 2	10	9
site 1	site 3	10	7
site 2	site 2	5	4
site 2	site 1	10	10
site 3	site 3	5	3
site 3	site 1	10	10

Table 4.7. Erosion values of gypsum blocks on the three test sites measured in the period 1989–1990, in g h^{-1} (n indicates the number of observations)

source	minimal value	mean value	maximal value	n
1	0.05	0.07	0.08	4
2	0.06	0.09	0.14	20
3	0.06	0.13	0.16	8

Table 4.8. Sedimentation loads and dominant sediment fractions on the three test sites measured in the period 1989–1990

source	daily sedimentation (g m^{-2} day^{-1})	dominant fractions (mm)
1	10 – 1160	0.09 – 0.30
2	30 – 1150	0.15 – 0.30
3	270 – 1450	< 0.09

Examples of transplanted sponges are shown in Fig. 4.12. Sponge A was transplanted from a sheltered site (site 1) to an exposed site (site 2). With sponge B the reverse experiment was done. The results of the experiments are summarized in diagrams as shown in Fig. 4.13. In this figure the resulting diagrams are shown for the transplanted sponges exhibited in Fig. 4.12. The circumference of the sponge at the beginning of the experiment is indicated with a dotted line (for non-marked branches only possible in the second experiment) and at the end as solid lines. The position of the needles, as detected from the x-ray photographs, is also visualized.

4.2.3 Comparison of Growth Forms of the Transplants and Simulation Experiments

In the diagrams (see Fig. 4.13) it can be seen that growth only takes place in certain parts of the sponges, viz. at the tips of the sponges; this is also demonstrated in the longitudinal section shown in Fig. 3.12. This is in agreement with the simulation models where growth only occurs at the apices. Only at the trunk of the sponge, close to the nail, is secondary growth observed. In photographs (see Fig. 4.12) it appears that the head of the nail is partly covered with sponge tissue. Also the attachment wire is incorporated in the sponge tissue. This secondary growth is probably

Fig. 4.12. Examples of transplanted *Haliclona oculata*. Sponge A was transplanted from the sheltered site 1 to the exposed site 2, with sponge B the reverse experiment was carried out (A 3.5 and B 1.5 month experiment).

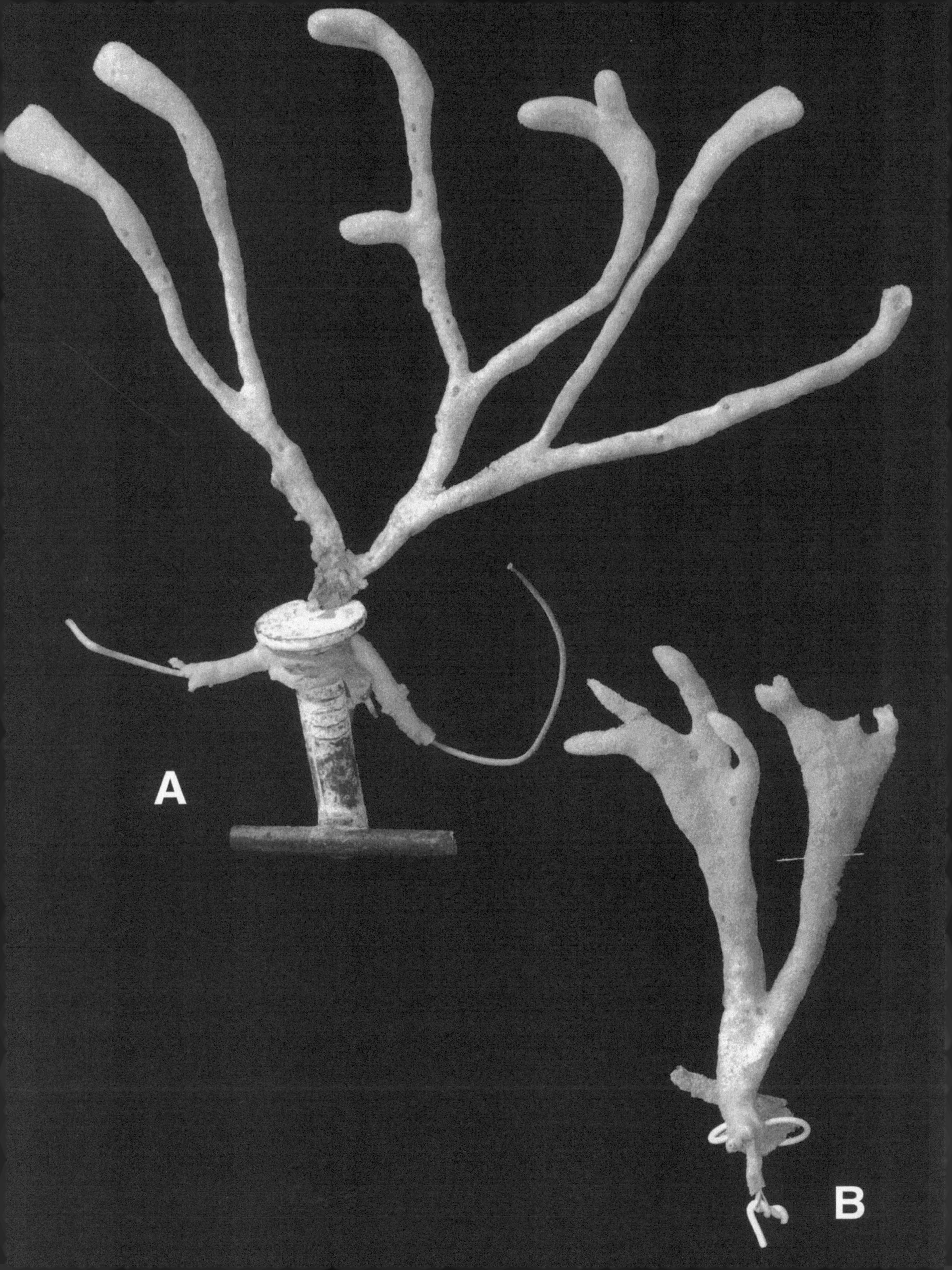
A
B

induced by local damage inflicted to the sponge when being removed from the substrate. The secondary growth can be introduced into the model by changing the status of non-active fertile elements (see Sect. 3.6.3) into active again. The consequence is that these tangential elements can participate again in the growth process.

The forms of the transplants before and after the experiment were compared by measuring the diameters of circles just fitting in the branches (see Fig. 4.14), comparable with the measurements done in Fig. 4.1. The contours shown in Fig. 4.13 were used for fitting circles within the branches.

Diameters of three types of circles were measured: the diameter da of the largest circle (a) which fits in the branch just before ramification, the diameter db of the largest circle (b) just after ramification, and the diameter dc of the largest circle c fitting in the top of a branch (see Fig. 4.14). The value of this last circle should vary between da and db. This assumption was verified by comparing the dc data sets with with the da, db data sets. The reason for using this measure instead of da or db is that only in long-lasting growth experiments are sufficient ramifications formed to allow enough measurements of da and db. Long-lasting experiments are less easy to interpret because of the increased chance of damage to the transplanted sponges in the course of time.

In order to detect the effect of the transplantations on the degree of plate formation at the extremities of the branches the resulting dc values are compared to a data set of dc values from the source site, measured

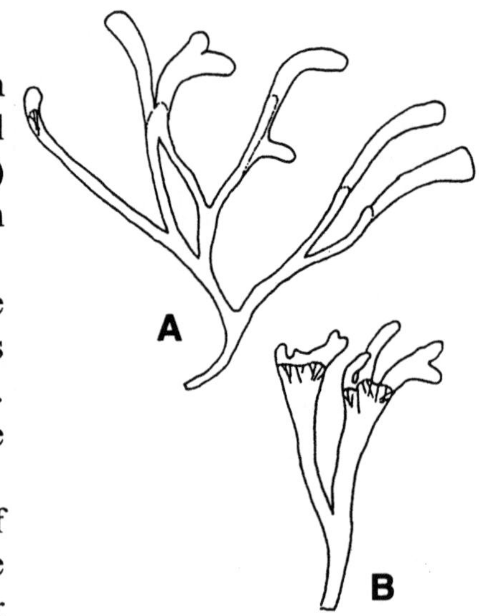

Fig. 4.13. Contours of sponges used for transplantation experiments. Sample A was transplanted from a sheltered site to an exposed site, with sample B the reverse experiment was carried out (A 3.5 and B 1.5 month experiment).

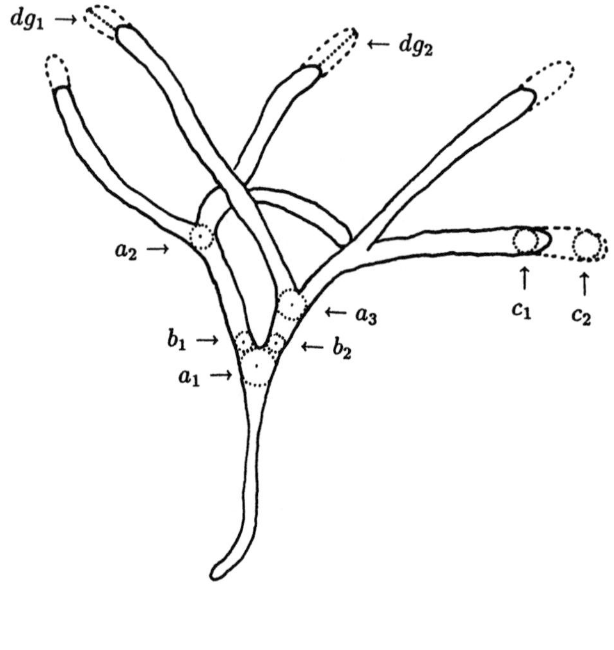

Fig. 4.14. Diagram of the contours of a sponge, with the locations of the measured circles. Circle a is the largest circle which fits in the branch just before ramification, b is the largest circle just after ramification, c is the largest one which fits in the top of a branch (da, db and dc are the corresponding diameters). dg is a measure of the amount of material added in the test period to the tip of the sponge.

in material collected at the three test sites. In the first experiment this data set was measured in the material collected additionally. In the second experiment this set was measured using the contour photographs of the transplants before the experiment. Additionally in the second experiment the amount dg of material added to the top of the sponge in the test period was measured. This distance was measured from the intersection point of the growth axis and the surface before the experiment (dotted line) and after the experiment.

Measuring the growth velocity

The distribution of dc

The assumption that the distribution of dc is situated between those of da and db was tested by applying the two-sample rank test for the material collected additionally at the three test sites. In all three cases the hypothesis that the distribution of dc is the same as da or db was tested against the alternative that they differ by a translation. In all cases a significance level of 5 % was applied. The results of this test are shown in Table 4.9.

Only in one case was this hypothesis accepted; in all other cases the assumption appeared to be correct. In Tables 4.10 and 4.11 the mean values of the three types of diameters for the three test sites are listed.

Table 4.9. Results of the two-sample rank test carried out to test the assumption that the distribution of dc is situated between those of db and da. This test was done for data sets collected in the additional material from the three test sites in the first experiment; in the second the measurements were made using the contours of the transplants before the experiment.

test site	type of diameter	result test (first experiment)	result test (second experiment)
1	da	dc left da	dc left da
1	db	dc right db	dc equals db
2	da	dc left da	dc left da
2	db	dc right db	dc right db
3	da	dc left da	dc left da
3	db	dc right db	dc right db

The values in Table 4.10 were measured in material collected additionally at the three test sites. The values in Table 4.11 were determined in the contours in the sponges before the experiment.

The changes in form in the transplants were determined by comparing the dc data sets from the additional material from the test sites to the dc data sets measured in the transplanted sponges. Both data sets were

Table 4.10. Mean values of the three types of diameters for the three test sites measured in the first experiment (n_x indicates the number of observations and s_x the standard deviations)

test site	$\overline{db}$	n_{db}	s_{db}	$\overline{dc}$	n_{dc}	s_{dc}	$\overline{da}$	n_{da}	s_{da}
1	0.15	102	0.04	0.16	51	0.03	0.26	102	0.05
2	0.20	131	0.06	0.24	50	0.07	0.36	131	0.08
3	0.19	46	0.04	0.21	28	0.04	0.32	24	0.05

Table 4.11. Mean values of the three types of diameters for the three test sites measured in the second experiment (n_x indicates the number of observations and s_x the standard deviations)

test site	$\overline{db}$	n_{db}	s_{db}	$\overline{dc}$	n_{dc}	s_{dc}	$\overline{da}$	n_{da}	s_{da}
1	0.15	52	0.05	0.15	47	0.03	0.24	44	0.07
2	0.20	52	0.05	0.24	50	0.08	0.33	52	0.06
3	0.18	43	0.04	0.18	51	0.04	0.29	41	0.08

compared by applying the two-sample rank test. The results of the comparison are summarized for both experiments in the Tables 4.12 and 4.13. In Table 4.14 the values of dg, i.e. the amount of material added to the tip of the sponge, measured in the transplants in the second experiment are shown.

Predicted forms

The model predicts that a sponge transplanted from a sheltered site to an exposed site will yield thin-branching sponges with plate-like ends (Fig 4.7A). The reverse experiment will yield palmate sponges with thin distal branches (Fig. 4.7B).

Forms in the transplantation experiments

The predicted forms were indeed found in the transplantation experiments. In Fig. 4.12A and Fig. 4.13A a thin-branching sponge is shown, which was transplanted from the more sheltered site 1 (mean erosion value 0.07 g h^{-1}, see Table 4.7) to the more exposed site 2 (mean erosion value 0.09 g h^{-1}), which exhibits plate-like extremities. In Fig. 4.12B and Fig. 4.13B a plate-like sponge is shown with which the reverse experiment was carried out and which depicts a palmate form with thin distal ends.

In general the shift from thin-branching to plate-like and the occurrence of palmate forms could be demonstrated by comparing the dc diameters of the sponges at the end of the transplantation experiment with the dc diameters at the beginning of the experiment. Table 4.12 shows that the value of dc, when sponges are transplanted from the sheltered

Table 4.12. Comparison of the dc (indicated as dc_o) data set from the additional material collected at source site with the dc (indicated as dc_r) data set measured in the transplants at the target site, for the experiment in the period 11 March 1989 until 30 April 1989 (n_x indicates the number of observations and s_x the standard deviations; the index o marks the measurements at the beginning and index r at the end of the experiment)

source	target	n_r	$\overline{dc_r}$	s_{dc_r}	result test	n_o	$\overline{dc_o}$	s_{dc_o}
1	1	3	0.18	0.03	r equals o	51	0.16	0.03
1	2	28	0.19	0.04	r right o	51	0.16	0.03
1	3	26	0.21	0.06	r right o	51	0.16	0.03
2	2	23	0.27	0.08	r equals o	50	0.24	0.07
2	1	36	0.20	0.07	r left o	50	0.24	0.07
3	3	26	0.29	0.06	r right o	28	0.21	0.04
3	1	17	0.21	0.06	r equals o	28	0.21	0.04

Table 4.13. Comparison of the dc (indicated as dc_o) data set from the contours of the transplants before the experiment with the dc (indicated as dc_r) data set measured in the transplants at the end of the experiment, in the period 30 November 1989 until 17 March 1990 (n_x indicates the number of observations and s_x the standard deviations; the index o marks the measurements at the beginning and index r at the end of the experiment)

source	target	n_r	$\overline{dc_r}$	s_{dc_r}	result test	n_o	$\overline{dc_o}$	s_{dc_o}
1	1	19	0.25	0.06	r right o	47	0.15	0.03
1	2	37	0.29	0.09	r right o	47	0.15	0.03
1	3	13	0.21	0.08	r right o	47	0.15	0.03
2	2	26	0.24	0.06	r equals o	50	0.24	0.08
2	1	57	0.28	0.08	r right o	50	0.24	0.08
3	3	–	–	–	–	51	0.18	0.04
3	1	43	0.25	0.09	r right o	51	0.18	0.04

site to the exposed site 2, increases. This indicates an increase of plate-formation comparable with the sponge depicted in Fig. 4.12A. For the reverse experiment a decrease of dc and a decrease of plate-formation is found, resulting in palmate objects. In the control experiments, in which the sponges were returned to the source site, it appears that the transplantation experiment itself does not result in a shift of the dc distribution for the sites 1 and 2.

In the second experiment (period 30 November 1989 until 17 March 1990) the shift from plate-like to thin-branching forms, resulting in palmate sponges, could not be demonstrated. In the experiments, carried out in the sites 1 and 2, a general increase in plate formation was found (see Table 4.13). This general increase of *dc* can be explained by the occurrence of heavy storms in 1990. Two severe storms struck the area (25 January 1990 and 26 February 1990) with wind speeds of 30 m s^{-1} and 26 m s^{-1} (averaged over one hour) respectively, which resulted in an overall increase in water movement at the test sites. The difference between the sheltered and the exposed site disappeared. Although this experiment did not demonstrate the emergence of palmate forms, it can explain that in some cases in the normally sheltered Lake Grevelingen (see Fig. 4.9), typically palmate sponges are found. An example of such a palmate form is exhibited in Fig. 4.15. During the occurrence of storms and the period afterwards in which the increase in water movement is gradually damped, the thin-branching sponges from this location are form plate-like ends. After the temporary increase in water movement the location becomes sheltered again. The plate-like ends presumably die off because the tissue is not in short-distance contact with the environment, or they gradually transform into palmate growth forms. This result is another indication that growth forms of sponges have a potential use as a continuous registration medium of environmental parameters.

Palmate sponges

In Table 4.14 the values of *dg*, the amount of material added to the top of the sponge, measured in the transplants in the second experiment can be compared for the three test sites. In general it is expected that the growth velocity will increase when the water movement and food supply increase on the test site. Because of the occurrence of stormy weather in the test period it was not possible to make a good distinction between an exposed and sheltered location: all locations were probably more or less exposed.

Growth velocity of the transplants

Table 4.14 shows that the transplants on site 3 exhibit a remarkably low growth velocity. In Tables 4.5 and 4.6 can be seen that for *Haliclona oculata* life is not very easy at site 3: there is a relatively low survival of the transplants at this site. In the second test period it was quite difficult to find enough living specimens to carry out the transplantation experiments. The deviating position of site 3 can be explained by its relatively high sediment load. Table 4.8 shows that the minimal sediment load is high compared to the other test sites (270 g m^{-2} day^{-1}), with bottom sediment (mainly consisting of fractions smaller than 0.09 mm).

With the simulation models it is possible to make some predictions about the effect of environmental changes on the growth forms. By com-

Fig. 4.15. Example of a palmate sponge, which is incidentally found in sheltered locations such as Lake Grevelingen (see Fig. 4.9)

Fig. 4.16. Example of a transplanted *Haliclona oculata* which was positioned horizontally at the nail (see Fig. 4.10) (3.5 month experiment)

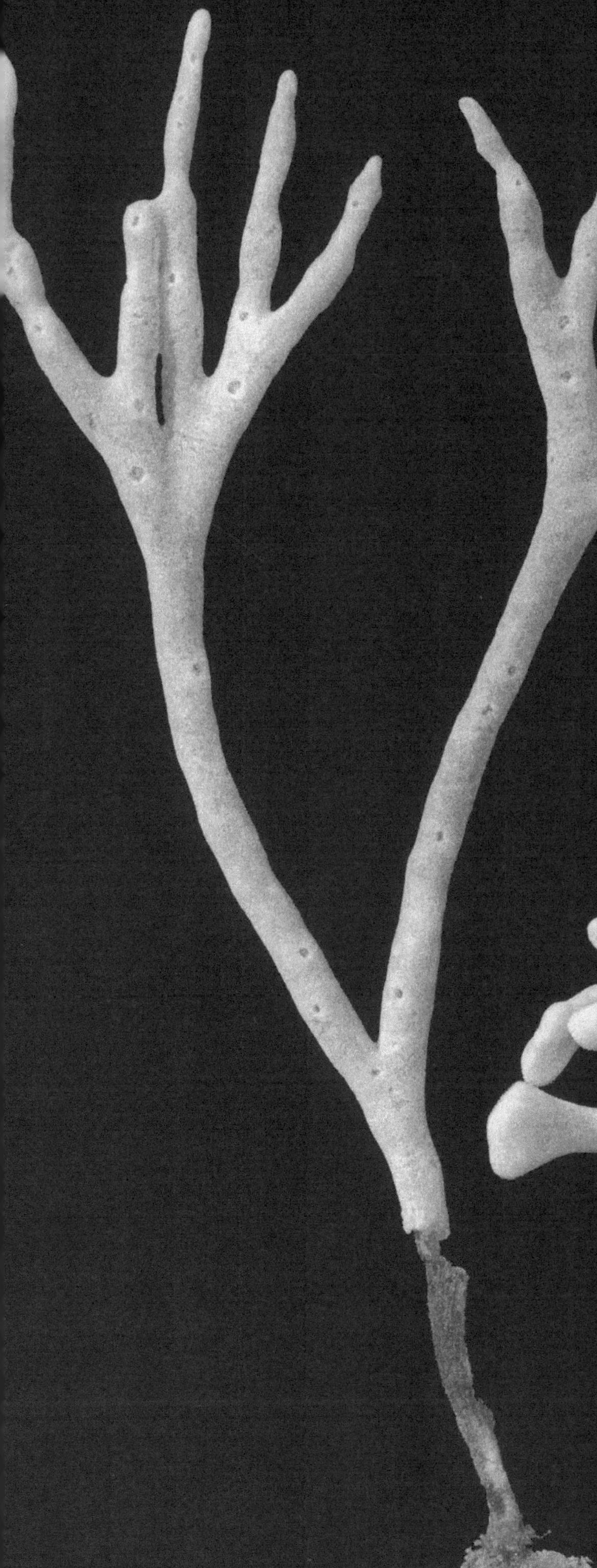

4.15

4.16

Table 4.14. Values of dg for the transplantation experiments done in the period 30 November 1989 until 17 March 1990 (n indicates the number of observations and s_{dg} the standard deviations)

source	target	n	$\overline{dg}$	s_{dg}
1	1	16	1.07	0.40
1	2	28	1.82	0.93
1	3	13	0.95	0.47
2	2	18	1.22	0.48
2	1	47	1.53	0.46
3	3	–	–	–
3	1	31	1.26	0.44

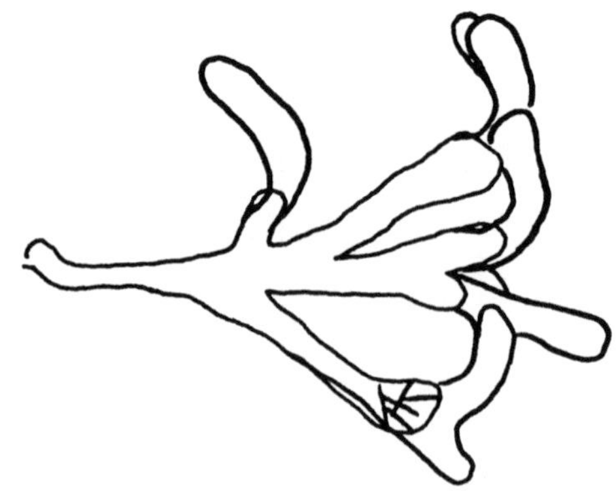

Fig. 4.17. Contours of the sponge *Haliclona oculata*, which was positioned horizontally and returned to its original growth site

paring growth forms from various sites and using simulation models, it becomes possible to detect sites with a deviating environmental condition.

In the second experiment a sponge was positioned horizontally at the nail (see Fig. 4.10). In Fig. 4.16 it can be observed from the attachment wire (the circle next to the holdfast of the sponge) that the sponge has been growing in a horizontal direction. There is a clear tendency in the sponge to grow upwards. Another example of a transplant which was positioned horizontally is shown in Fig. 4.17. In this figure the contours of the sponge (taken from photographs) at the beginning and at the end of the experiment are indicated. This effect shows that the model in Fig. 3.17J can be used to predict forms which are found in experiments where a sponge was positioned horizontally (compare Fig. 4.8). It is also experimental evidence for the negative substrate-tropism assumed in the model Fig. 3.17J.

4.3 Conclusions

Figure 4.4 and Table 4.3 show that plate-formation (resulting in a relatively higher value for da and db) and branching (a lower value for rb) increases with water movement. This example shows that growth forms can be used for bio-monitoring purposes: from the actual forms it is possible to derive some of the environmental conditions in which the form emerged. In a range with increasing water movement, the irregularity of the objects increases, resulting in a higher fractal dimension (see Table 4.4). All these tendencies can also be demonstrated in simulation experiments (see Tables 4.1 and 4.4 and Fig. 4.3 and 4.6) in which a range of forms was generated where the parameters *lowest_value* and *max_curv* change simultaneously. From Fig 4.3 the relation between an observed form and the model parameter *max_curv* can be derived. In theory a model could be generated with a high correspondence to the actual samples by tuning this parameter.

The predictions made by the model (Fig. 4.7A), that a sponge transplanted from a sheltered site to an exposed site will yield thin-branching forms with plate-like ends (Fig 4.7A) while the reverse experiment will yield palmate sponges with thin distal branches (Fig. 4.7B), could also be demonstrated in reality. This effect could be demonstrated in many transplants (see Table 4.12). Also the predicted effect on the growth form (see Fig. 4.8), when the transplant is rotated and positioned horizontally during the growth process, could be verified in an experiment shown in Fig. 4.17.

5 3D Models of Growth Forms

In this chapter the development of a model of a growth in three dimensions is discussed. In the first section it is demonstrated how the modelling system for iterative geometric constructions (see also Sect. 2.6) can be extended to 3D. In Sect. 5.2, a discussion follows on the 3D structure of an organism with radiate accretive growth. In Sects. 5.3 and 5.4 it is discussed how this 3D structure can be represented in a model. The results of these sections, a suitable data representation for a 3D object developing in the radiate accretive growth process, is used in the final Sect. 5.6, in which the development of a model of a radiate growth process in three dimensions is presented. Most of the rules discussed in Sect. 3.6, on the iterative geometric constructions for simulating this growth process, will be extended to 3D. In Sect. 3.6 the biological examples used as a case-study were the sponge *Haliclona oculata* and the stony coral *Montastrea annularis*. For reasons which will become clear in the next sections, the sponge *Haliclona simulans* (see Fig. 3.15) is used as an example in this chapter. The biological significance of the rules will be indicated only briefly in this chapter, since most of them were already discussed in Sect. 3.6. The extension to 3D of the simulation model is an essential one, since many aspects of the growth process (e.g. a larger possibility for the branches to avoid each other, the formation of flattened forms influenced by the flow direction) can only be adequately described with a 3D model. For convenience the symbols used in the sections on the 3D model of radiate accretive growth (Sects. 5.3–5.7 are listed separately in Sect. 5.8).

5.1 Constructions in Space, a 3D Modelling System for Iterative Constructions

Some of the 3D equivalents of the classical fractal objects can be generated with a relatively simple extension of the 2D modelling system described in

Sect. 2.6. For this purpose the rules for representing the three components in the production rule should be extended with another component: the *face*. For a 3D construction it is necessary to define from which type of surface the three previous components are constructed. The base elements are now built themselves from lower level elements, the *faces*. Each *face* contains the references to the vertices. The vertices are stored separately in a list in order to avoid an enormous amount of overhead, because the same vertex occurs several times in various surfaces.

Face

The algorithm of the 2D modelling system (see Sect. 2.6.2) is in (a simple version of) the 3D system extended by a new level in G in which the vertices are taken from the *faces*. The transformation in G1 (see (2.15) is changed into a 3D transformation:

$$\begin{aligned} M_{i,j} &= R_{rp,axis}(\gamma).S(sf,sf,sf). \\ &\quad T(VG_{xk}-VG_{x0},VG_{yk}-VG_{y0},VG_{zk}-VG_{z0}) \\ &\quad \text{for } 0 \leq k \leq m \end{aligned} \tag{5.1}$$

The transformation is a combination of a rotation (R, with rotation point rp and $axis$ as axis of rotation), scaling (S, with scaling factor sf, see (2.14)), and a translation (T, the translation vector is determined by analogy with (2.15)). The visualization in the modelling and the interactive design of the four components in the production rules are less trivial for the 3D objects (on 3D visualization techniques see Newman and Sproull 1979; Foley et al. 1990).

With the 3D modelling system some of the 3D equivalents of the curves shown in Sect. 2.6 can be generated. The construction of a 3D equivalent of the Sierpinski arrowhead (Fig. 2.30) is shown in Fig. 5.1. The first three components consist of triangles and the first two components are constructed from the third component: a tetrahedron. In the construction

Sierpinski arrowhead

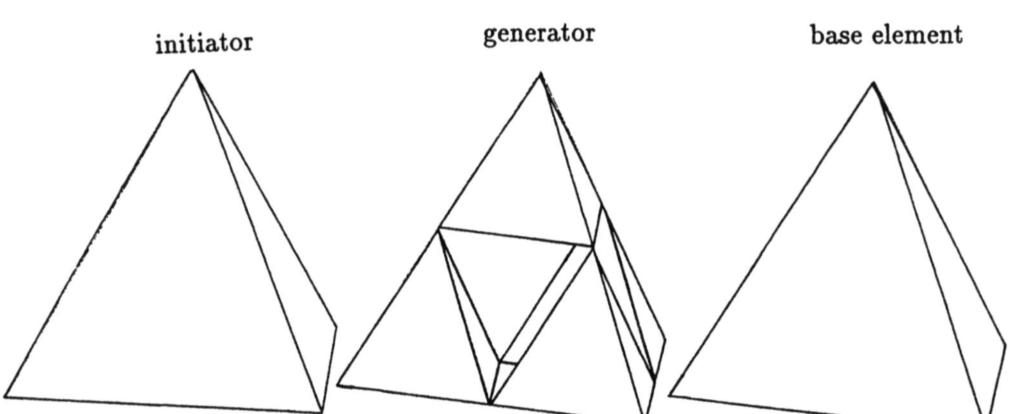

Fig. 5.1. Construction of a 3D equivalent of the Sierpinski arrowhead, with the result shown in Fig. 5.2

each tetrahedron is replaced by a set of four tetrahedra in each iteration step. The result of this construction is shown in Fig. 5.2. The replacement system is shown below:

$$
\begin{aligned}
face &= (V_p, V_q, V_r) \qquad (5.2)\\
base & \\
element &= (face_a(V_a, V_b, V_c)); face_b(V_a, V_c, V_d); face_c(V_a, V_d, V_b);\\
& \quad face_d(V_b, V_d, V_c));\\
initiator &= be((face_a(V_0, V_1, V_2)); face_b(V_0, V_2, V_3); face(V_0, V_3, V_1);\\
& \quad face(V_1, V_3, V_2));\\
generator &= be((face_a(V_{4*i}, V_{4*i+1}, V_{4*i+2})); face_b(V_{4*i}, V_{4*i+2}, V_{4*i+3});\\
& \quad face(V_{4*i}, V_{4*i+3}, V_{4*i+1}); face(V_{4*i+1}, V_{4*i+3}, V_{4*i+2})); \rightarrow\\
& \quad be((face_a(V_{4*i}, T_{1j}(V_{4*i}), T_{2j}(V_{4*i}));\\
& \quad face_b(V_{4*i}, T_{2j}(V_{4*i}), T_{3j}(V_{4*i}));\\
& \quad face(V_{4*i}, T_{3j}(V_{4*i}), T_{1j}(V_{4*i}));\\
& \quad face(T_{1j}(V_{4*i}), T_{2j}(V_{4*i}), T_{3j}(V_{4*i})));\\
& \quad be((face_a(T_{1j}(V_{4*i}), V_{4*i+1}, T_{4j}(V_{4*i}));\\
& \quad face_b(V_{4*i}, T_{4j}(V_{4*i}), T_{5j}(V_{4*i}));\\
& \quad face(T_{1j}(V_{4*i}), T_{5j}(V_{4*i}), V_{4*i+1});\\
& \quad face(V_{4*i+1}, T_{5j}(V_{4*i}), T_{4j}(V_{4*i})));\\
& \quad be((face_a(T_{2j}(V_{4*i}), T_{4j}(V_{4*i}), V_{4*i+2});\\
& \quad face_b(T_{2j}(V_{4*i}), V_{4*i}2, V_{4*i+2});\\
& \quad face(T_{2j}(V_{4*i}), T_{6j}(V_{4*i}), T_{4j}(V_{4*i}));\\
& \quad face(T_{4j}(V_{4*i}), T_{6j}(V_{4*i}), V_{4*i+2}));\\
& \quad be((face_a(T_{3j}(V_{4*i}), T_{5j}(V_{4*i}), T_{6j}(V_{4*i}));\\
& \quad face_b(T_{3j}(V_{4*i}), T_{6j}(V_{4*i}), V_{4*i+3});\\
& \quad face(T_{3j}(V_{4*i}), V_{4*i+3}, T_{5j}(V_{4*i}));\\
& \quad face(T_{5j}(V_{4*i}), V_{4*i+3}, T_{6j}(V_{4*i})));
\end{aligned}
$$

Seeding square (3D equivalent)

The construction of a 3D equivalent of the seeding square (see Fig. 2.25) is shown in Fig. 5.3. In this construction in each iteration step a square is replaced by a set of 15 new squares, with the result shown in Fig. 5.4. The construction of an infinitely empty cube (see Fig. 5.6, the Menger sponge; see also Mandelbrot (1983)) is shown in Fig. 5.5. In this construction a cube is replaced by a set of 20 cubes.

Menger sponge

In this version of the 3D modelling system several disadvantages occur:

Fig. 5.2. 3D equivalent of the Sierpinski arrowhead constructed with the rule shown Fig. 5.1

a) In the 2D system it is trivial to identify which edges border a certain edge in a (connected) curve of edges. In the 3D system, it is quite difficult to detect which faces are neighbouring to a certain face in a connected mesh of faces. A face which is neighbouring to another face has two (references to) vertices in common with the other face. Especially for objects constructed from a large number of vertices, this test will be rather time-consuming. In the section on 3D models for modelling growth processes, it will appear that this test is a crucial one.
b) The system is less general, and it is not possible to develop 3D equivalents for all curves and objects shown in the section on the 2D modelling system. An example is the construction of a 3D equivalent of the ramifying fractals (see Fig. 2.23). A possible solution could be to extend the production rules to rules where several types of surfaces are used. A possible construction for a 3D equivalent of the ramifying objects is displayed in Fig. 5.7. In this extended rule the three components are constructed of tubes, in which as a *base element* a tube is used consisting of a mixture of two types of surfaces: n rectangular surfaces (A) and an n-sided circular surface (B). Examples of objects generated with this type of rule are shown in Kawaguchi (1982).

The problems shown in this section will return in Sect. 5.6 on 3D models of radiate accretive growth, where a solution is suggested for point a).

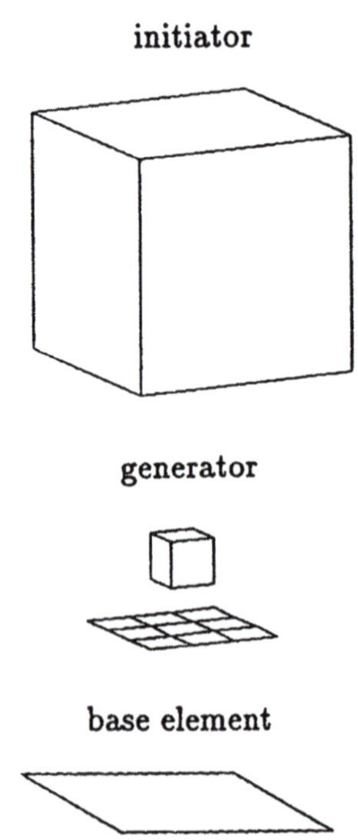

Fig. 5.3. The construction of a 3D equivalent of the seeding square (see Fig. 2.25), with the result shown in Fig. 5.4

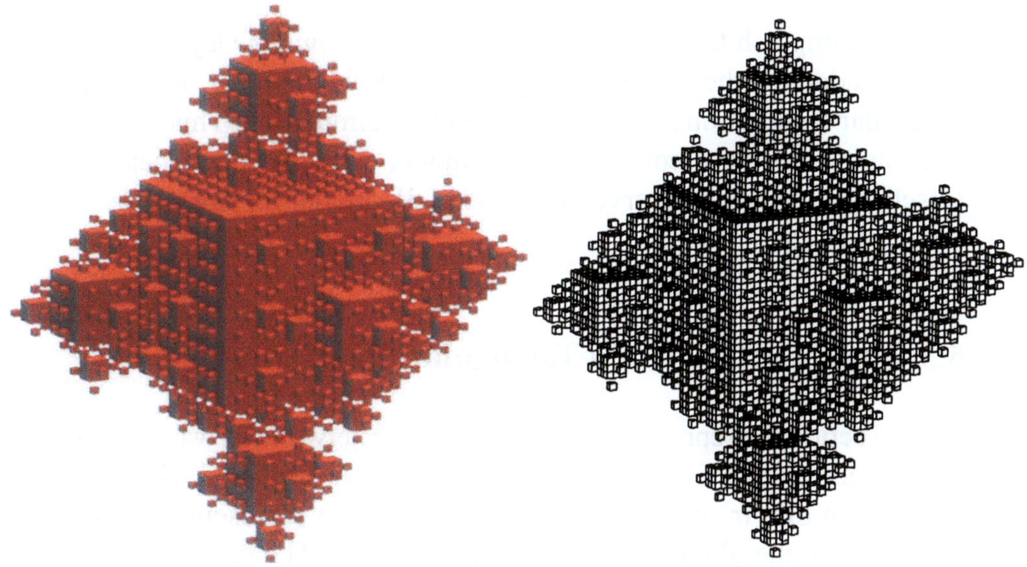

Fig. 5.4. 3D equivalent of the seeding square constructed with the rule shown Fig. 5.3

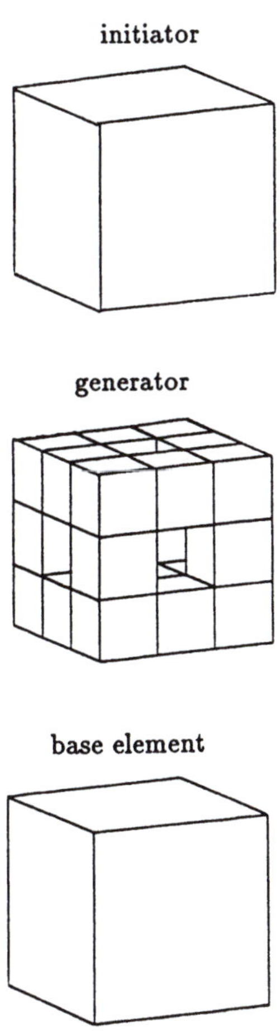

Fig. 5.5. The construction of the Menger sponge (see Fig. 5.6)

5.2 Description of an Organism with Radiate Accretive Growth and a Triangular Tessellation of the Surface

The sponge *Haliclona simulans* (see Fig. 3.15) is closely related to *Haliclona oculata*; the internal architecture of the latter is discussed in Sect. 3.5. In *Haliclona oculata* the tangential elements are organized in 4- to 6-sided polygons (see Fig. 3.13A). In a tangential microscopic view of *Haliclona simulans* (see Fig. 5.8) it can be seen that the tangential spicula are arranged in a triangular mesh. This triangular arrangement is characteristic for several species in the sponge family Chalinidae (see De Weerdt 1986). The longitudinal tracts are arranged in about the same way as in *Haliclona oculata* (see Fig. 3.13B). The longitudinal bundles are somewhat larger in diameter and more than 2 spicula thick. In *Haliclona simulans* the longitudinal bundles may vary in length and the 3D mesh of spicula shows a radiate symmetry, in correspondence with the structure of *Haliclona oculata*.

The triangles in the tangential view in Fig. 5.8 can be thought to be arranged in, predominantly, pentagons and hexagons. In some cases 7-8 spicula meet in one point and heptagons and octagons are formed. In an idealized version of the triangular network, it can be transformed into a network of 5- to 6-sided polygons, when some of the spicula which meet in one point are removed. The result of this transformation is a *Haliclona oculata*-like network. This difference together with a more extensive aquiferous system in *Haliclona simulans* (for a comparison with *Haliclona oculata* see Sect. 3.5) are the main architectural differences which distinguish the two species. In *Haliclona simulans* layers of triangles are added to the preceding stages in the growth process. This layered triangular tessellation is a suitable subject for simulation. This triangular tessellation can be used as a basis to model a layered penta-hexagonal tessellation, which can serve as a general 3D model for organisms with radiate accretive growth.

5.3 Representation of a Triangular Tessellation

In this section the representation of a triangular network, as was described in the preceding section for the biological object, is discussed. This formal description of a triangular network is necessary to represent the surface of objects in the 3D model of the growth process. The main goal in this section is to develop a method for the representation of a surface which is tiled with discrete skeleton elements, which can only vary somewhat in size. The skeleton elements are represented as edges of triangles and these should be as equilateral as possible. With the created representation it should be possible to tessellate the convex and concave surfaces which emerge in the growth process.

A triangular mesh can be represented by a vertex index list $\mathcal{V}$:

$$\mathcal{V} = (V_1, ..., V_n) \tag{5.3}$$

in which each index refers to a coordinate triple in the list $\mathcal{C}$:

$$\mathcal{C} = ((x_1, y_1, z_1), ..., (x_n, y_n, z_n)) \tag{5.4}$$

and a list $\mathcal{T}$ of triangles:

$$\mathcal{T} = (T_1, ..., T_m) \tag{5.5}$$

Each triangle T_i from the list $\mathcal{T}$ consists of a triple of indices:

$$\begin{aligned} T_i &= (V_a, V_b, V_c) \\ & \quad V_a, V_b, V_c \in \mathcal{V} \end{aligned} \tag{5.6}$$

In a triangular tessellation of a surface each triangle is surrounded by 3 neighbouring triangles. In the representation of (5.6), the neighbouring triangles can only be identified by comparing all vertex indices of the triangle T_i with the indices of the other triangles from the list $\mathcal{T}$.

A more convenient representation of a triangle, where its neighbours can be identified straightforwardly, is given in the representation of T_i in (5.7). In this representation also the indices of the neighbouring triangles are included:

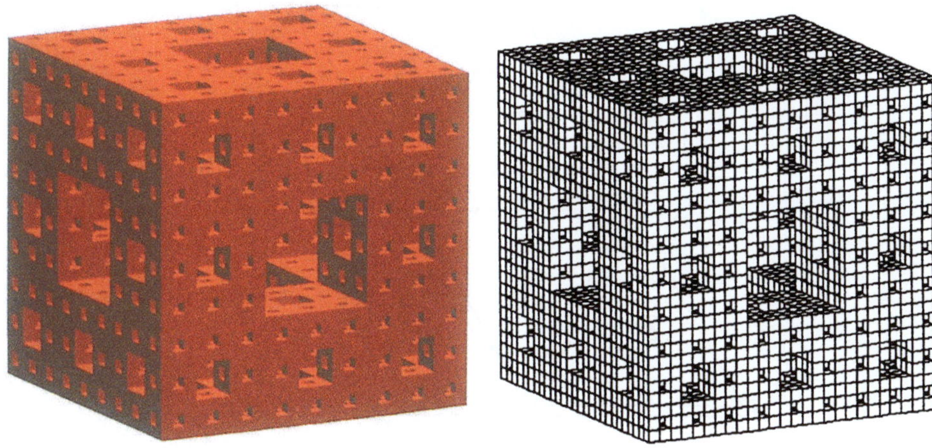

Fig. 5.6. The Menger sponge constructed with the rule shown in Fig. 5.5

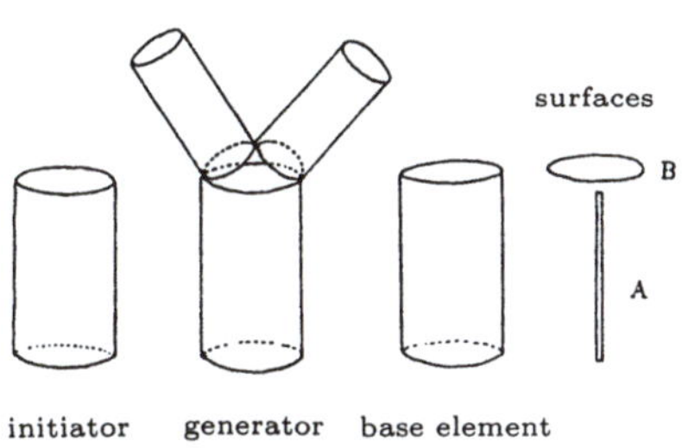

Fig. 5.7. A 3D construction of a ramifying fractal, where a mixture of two types of surfaces is used in the construction of the base element

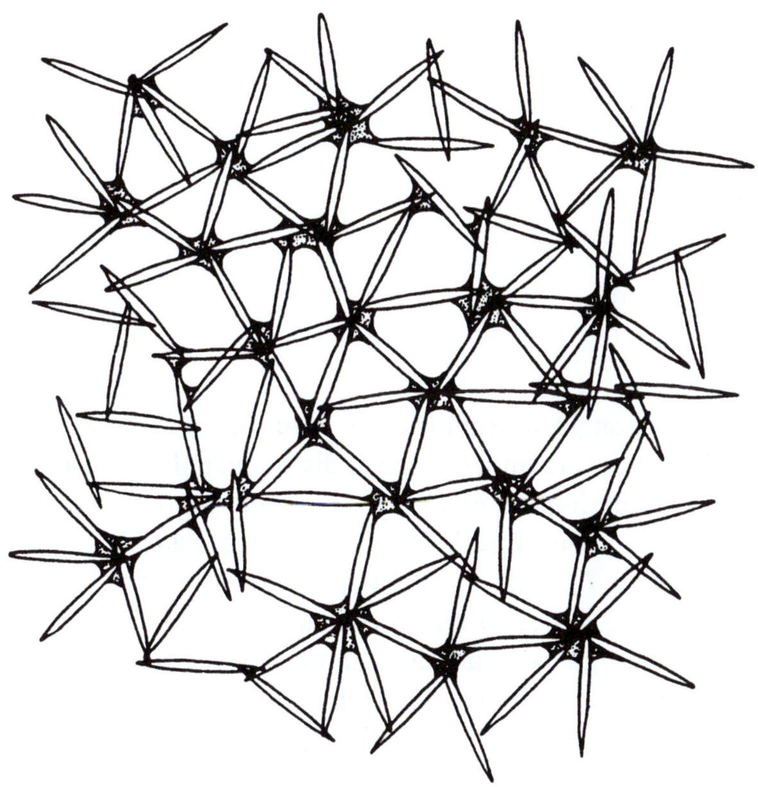

Fig. 5.8. Tangential view of the arrangement of spicula in the sponge *Haliclona simulans* (after De Weerdt 1986)

$$
\begin{aligned}
T_i &= ((V_a, V_b, V_c), (T_a, T_b, T_c)) \\
& V_a, V_b, V_c \in \mathcal{V} \\
& \text{triangle } T_a \text{ borders at } edge(V_a, V_b) \\
& \text{triangle } T_b \text{ borders at } edge(V_b, V_c) \\
& \text{triangle } T_c \text{ borders at } edge(V_c, V_a) \\
& T_a, T_b, T_c \in \mathcal{T}
\end{aligned} \qquad (5.7)
$$

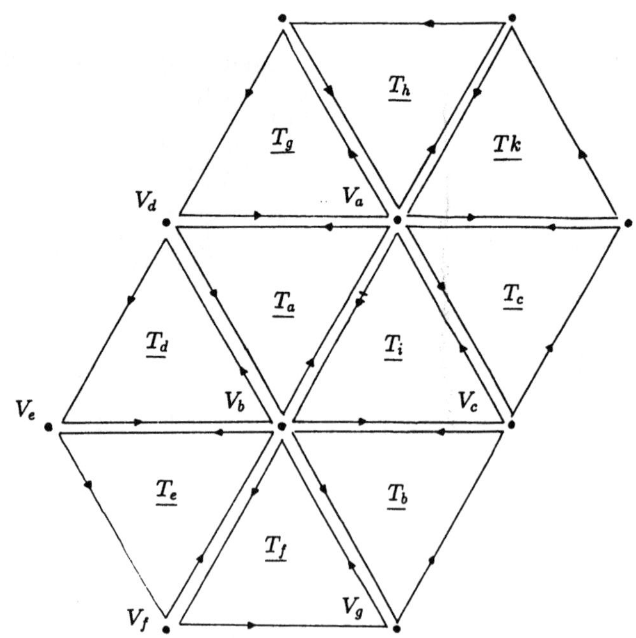

Fig. 5.9. Diagrammatic description of the triangle T_i from (5.7). The indices of the surfaces are underscored.

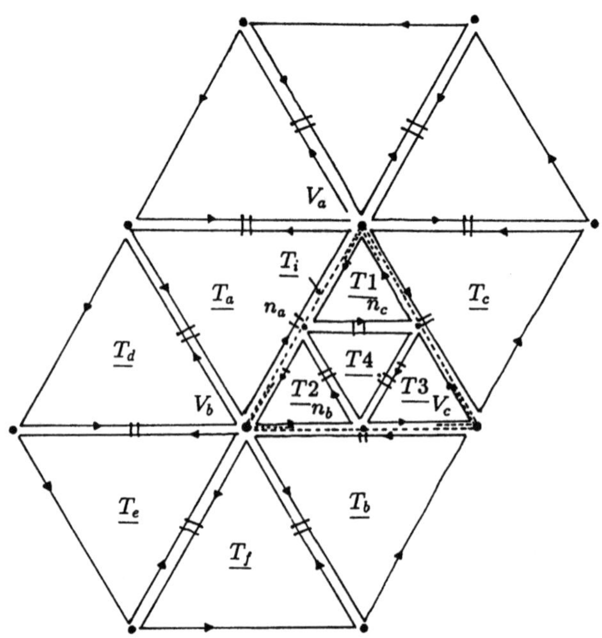

Fig. 5.10. Diagrammatic description of the triangle T_i after subdivision into four children $T1..T4$. The representation of the parent and children triangles are given in (5.11) and (5.10). The connections between triangles are indicated with the "=" symbol. A parent triangle is indicated as a dotted line grouping two children triangles together.

In Fig. 5.9 the triangle T_i from (5.7) is described in a diagram. The indices of the surfaces are underscored to distinguish them from the vertex indices. The vertex indices of triangle T_a bordering on T_i are arranged oppositely to the vertices in T_i. There are three possible arrangements for the vertices in T_a, viz.: V_b, V_a, V_d; V_a, V_d, V_b and V_d, V_b, V_a. The arrangement of the vertices is indicated in the diagram with an arrow. The first edge of a triangle is displayed with an arrow and a line. By consistently doing so, it is possible to define the normal vector of each triangle unambiguously (Möbius rule). When T_a, T_b or T_c equals $null$ in (5.7), the edge (V_a, V_b), (V_b, V_c) or (V_c, V_a) does not border on another triangle. In the representation of (5.7) it is easy to identify in a mesh of triangles the neighbours of a triangle T_i and also to determine the set of triangles which surround a vertex of the triangle T_i. The set of triangles surrounding the vertex V_b can be denoted as $set\_triangles(T_i, V_b)$, the set of vertices surrounding the vertex V_b as $set\_vertices(T_i, V_b)$ (see (5.8)). These vertices are only accessible using the list of triangles T, and for this reason the triangle T_i is also used as an argument in both sets.

$$set\_triangles(T_i, V_b) = \{T_i, T_a, T_d, T_e, T_f, T_b\} \qquad (5.8)$$
$$set\_vertices(T_i, V_b) = \{V_b, V_a, V_c, V_d, V_e, V_f, V_g\}$$

In the representation of (5.7) it is not possible to subdivide a triangle T_i into new triangles while leaving the neighbours unchanged. The necessity of doing this occurs when an additional condition (5.9) is applied which states that the edges of T_i may only vary between two limits. This condition stipulates that the triangles in the tessellation should remain as equilateral as possible. This property is disturbed when the tessellation is used to cover a convex or concave surface and can be restored by subdividing the non-fitting triangles. In an actual object one can imagine that the surface is tessellated with discrete-sized elements. During the growth process the surface is enlarged and new building elements have to be inserted in order to preserve the coherence in the tessellation. The limits depend on the size s, which is a constant describing the basic size of a tangential element in the mesh. The limits describe the degree of non-equilaterality of triangles; the values of the limits were chosen arbitrarily.

Basic size of a tangential element

$$\begin{aligned} 0.5s \leq \|V_a, V_b\| \leq 1.5s \text{ and} \\ 0.5s \leq \|V_b, V_c\| \leq 1.5s \text{ and} \\ 0.5s \leq \|V_c, V_a\| \leq 1.5s \end{aligned} \qquad (5.9)$$

In the case that T_i is enlarged and all edges exceed $1.5s$, it is necessary to subdivide T_i into four new triangles and a situation as described in Fig. 5.10 might occur. The parent triangle T_i is split up into four children $T1..T4$ and three new points n_a, n_b, n_c are added to the vertex list $\mathcal{V}$. The three new vertices n_a, n_b and n_c are situated at the middle of the subdivided edges.

An extension of the representation in (5.7), where triangles can be subdivided while leaving the neighbour triangles unchanged, is shown below:

$$T_i = \begin{cases} (\{\text{set of 3 vertex indices}\}, \{\text{set of 1-3 neighbour triangles}\}, \\ \{\text{set of 0-2 parents}\}, level) \\ \text{for } level = 0 \\ (\{\text{set of 2 children and 1 } null\}, \\ \{\text{set of 1 neighbour and 2 } nulls\}, \\ \{\text{set of 0-2 parents}\}, level) \\ \text{for } level > 0 \end{cases}$$

Hierarchical description of a triangular mesh

With this representation it is possible to create a hierarchical description of a triangular mesh in which a parent triangle can be subdivided into children triangles.

A representation of a child $T1$ is:

$$\begin{aligned} T1 &= ((V_a, n_a, n_c), (null, T4, null), (T_i, T_j, 0)) \quad (5.10)\\ & V_a, n_a, n_c \in \mathcal{V}\\ & \text{triangle } T4 \text{ borders at } edge(n_a, n_c)\\ & T4, T_i, T_j \in \mathcal{T} \end{aligned}$$

The triangle $T1$ shares an edge with the triangle $T4$ but the other edges are shared with no other triangles. The nonexistent neighbours are indicated with the value *null* in the second triplet. In the description of $T1$ a reference to its parent T_i is included, and the very last number indicates that $T1$ is a 0-level triangle, which is not further subdivided.

Level of subdivision

A representation of the parent T_i is:

$$\begin{aligned} T_i &= ((T1, T2, null), (T_a, null, null), (null, null, 1)) \quad (5.11)\\ & \text{triangle } T_a \text{ borders at } edge(V_a, V_c)\\ & T1, T2, T_a \in \mathcal{T} \end{aligned}$$

The indices in the first triplet indicate that T_i is composed of two triangles $T1$ and $T2$. The second triplet shows that the parent triangle T_i borders on the triangle T_a. The last triplet indicates that the parent T_i is not enclosed in any other parent triangle and that T_i is only subdivided once (level 1). In the representation of the child triangle (5.10) it can be seen that the level 0 triangle $T1$ is enclosed in two level 1 parent triangles T_i and T_j. T_j is described as:

$$\begin{aligned} T_j &= ((T1, null, T3), (null, null, T_c), (null, null, 1)) \quad (5.12)\\ & \text{triangle } T_c \text{ borders at } edge(V_c, V_a)\\ & T1, T3, T_c \in \mathcal{T} \end{aligned}$$

In this representation a parent triangle consists of two children and one *null* reference. The children in the parent triangles are ordered in a way corresponding to the vertices V_a, V_b, V_c of the triangle T_i before the subdivision. The children triangles $T1..T4$ are enclosed in three parent triangles (T_i, T_l, T_j). The ordering in the first triplet for these parent triangles is respectively: $(T1, T2, null)$, $(null, T2, T3)$, $(T1, null, T3)$.

Connections between triangles

In the diagram of Fig. 5.10 the connections between triangles are indicated with the "=" symbol. A parent triangle is indicated as a dotted line grouping two children triangles together. The edges of triangles which are at the border of the tessellation and which do not neighbour other triangles are unconnected. There can only exist a connection between bordering triangles when they share the same edge. In the diagram it can be seen that there is a connection between the triangle T_a:

$$\begin{aligned} T_a &= ((V_b, V_a, V_d), (T_i, T_g, T_d), (null, null, 0)) \qquad (5.13) \\ & V_b, V_a, V_d \in \mathcal{V} \\ & T_i, T_g, T_d \in \mathcal{T} \end{aligned}$$

Set of triangles surrounding a vertex

and T_i. From T_i (5.11) it is possible to identify the two children $T1$ and $T2$, and from both it can be derived that they border to $T4$. In this representation it is again possible to identify the neighbours of a child triangle $T1$ and to determine the set of triangles which surround a vertex of $T1$. For example:

$$\begin{aligned} set\_triangles(T1, V_a) &= \{T1, T_a, T_g, T_h, T_k, T_c\} \qquad (5.14) \\ set\_triangles(T1, n_a) &= \{T1, T_a, T2, T4\} \end{aligned}$$

Transition from large triangles into small ones

In the subdivision of T_i there are four children $T1..T4$ and two parents T_l, T_j introduced. T_i is changed from a level 0 triangle into a level 1 triangle. With this representation, where a parent contains only two children instead of three or four, a mesh as shown in Fig. 5.11 can also easily be described. In this mesh, where a transition from large triangles into small ones occurs, only two triangles of level 1 are necessary to describe the situation.

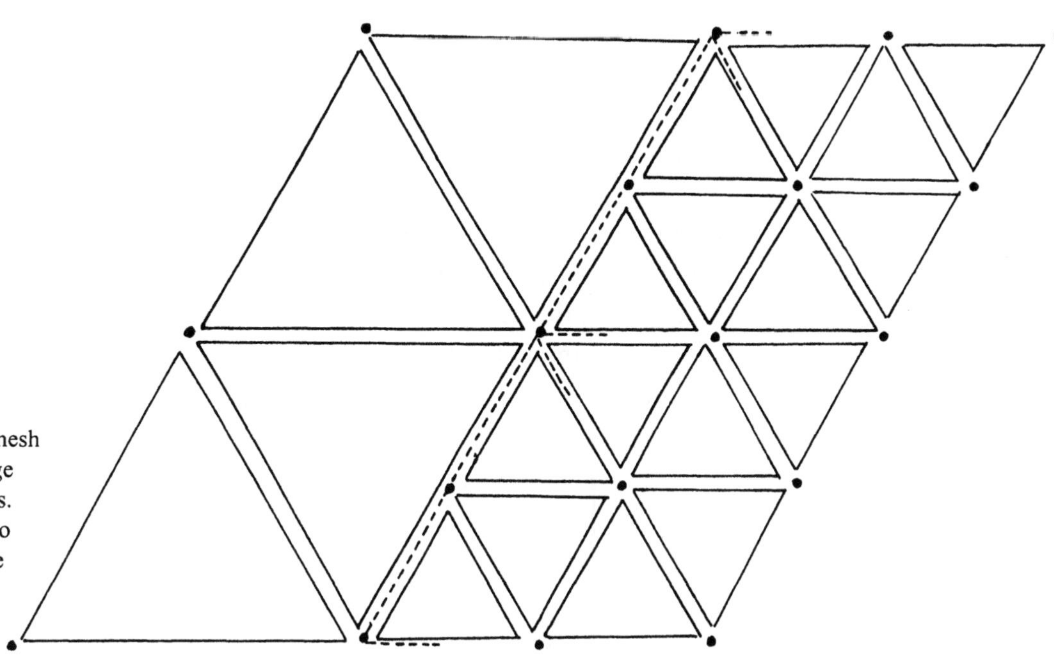

Fig. 5.11. Diagrammatic description of a triangular mesh where a transition from large into smaller triangles occurs. To describe the situation two level 1 (parent) triangles are necessary.

With the representation in (5.10) it is possible to create a hierarchical description of a triangular mesh, where a triangle can be n times subdivided. The vertex and triangle indices of the parent T_i can be arranged in a hierarchical structure, as depicted in Fig. 5.12, where the vertices are the leaves of the tree-structure (see also De Floriani 1989). Between the 0-level triangles $T1$ and T_a there is a difference of one subdivision. With the representation in (5.10) it is possible to create a triangular mesh, arranged hierarchically, where the difference in number of subdivisions between bordering triangles is higher than one. This type of mesh will not satisfy the rule (5.9), where the edges of each triangle in the mesh may only vary between two limits. In the examples of triangular meshes shown in the next sections, the maximal difference in number of subdivisions between bordering triangles will be one. In the representation of (5.10) it is also possible to create a mesh with "unnecessary" parent triangles (see Fig. 5.13). For simplicity reasons these superfluous parent triangles are removed from the list of triangles in the examples of triangular tessellations discussed later on. As a consequence parent triangles will only occur in transition zones as displayed in Fig. 5.11. As will be demonstrated in Sect. 5.6, it is possible to cover a 3D curved surface with triangles which can be considered as built from discrete skeleton elements and where the triangles remain almost equilateral.

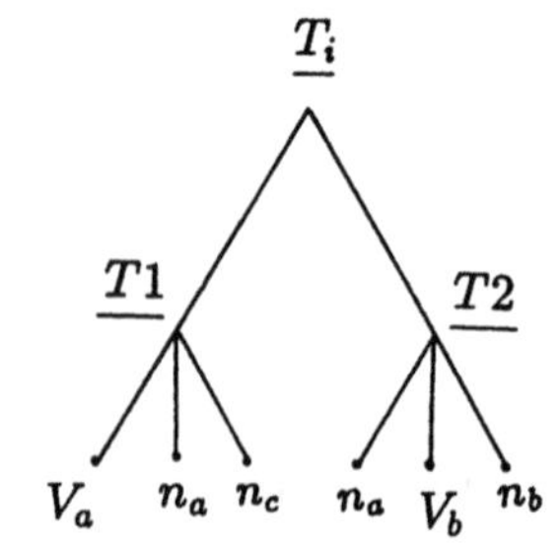

Fig. 5.12. Hierarchical structure of the vertex and triangle indices of the parent T_i from Fig. 5.10. The vertex indices are the leaves of the tree.

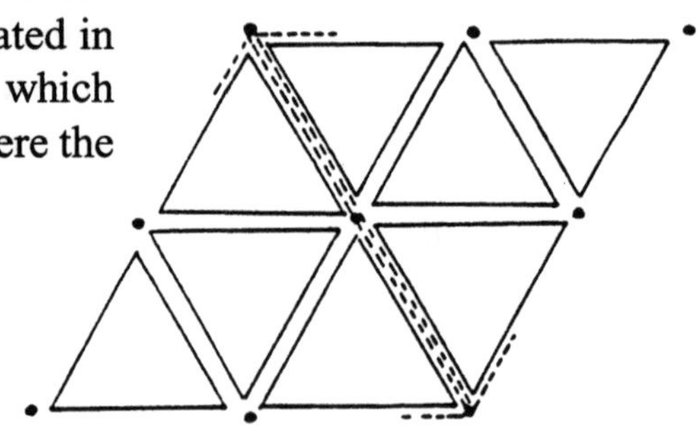

Fig. 5.13. Triangular mesh with two "unnecessary" parent triangles which can be replaced by normal connections

5.4 Representation of a Multi-Layer Triangular Tessellation

In the previous section a description was given of a possible representation of a triangular tessellation covering the surface of a 3D object. In this section the representation is extended to a multi-layer system. This extension is necessary, since in the simulation of the growth process in Sect. 5.6 layers consisting of triangles are constructed on top of each other in the succeeding growth steps. An example of a layered system, where a triangular tessellation is constructed upon another triangular tessellation, is shown in Fig. 5.14. The edges of the triangles are situated at the surface of the layers. These edges will be indicated as the tangential elements of the layered structure. The indices of the vertices can be represented as a list of vertex index lists:

$$\mathcal{V} = \begin{cases} V_{1,1}, V_{2,1}, \ldots \\ V_{1,2}, V_{2,2}, \ldots \\ . \\ . \\ V_{1,m}, V_{2,m}, \ldots \end{cases} \tag{5.15}$$

where each index refers to a corresponding list of coordinate lists. The tessellated layers can be represented in a list of triangle lists:

$$\mathcal{T} = \begin{cases} T_{1,1}, T_{2,1}, \ldots \\ T_{1,2}, T_{2,2}, \ldots \\ . \\ . \\ T_{1,m}, T_{2,m}, \ldots \end{cases} \tag{5.16}$$

A triangle list $T_{1,j}, T_{2,j}, \ldots$ in this list will be indicated as $layer(j)$. Each triangle $T_{i,j}$ from $\mathcal{T}$ can be represented in the form of (5.10).

In Fig. 5.14 $layer(3)$ can be considered as constructed upon $layer(2)$ and $layer(2)$ upon $layer(1)$. In the same sequence the total number of vertex indices, for each layer, increases when the surface of new layer is larger than the previous one. Between some of the vertices there is a straightforward correspondence. For example, the vertices $V_{a,3}$, $V_{b,3}$ and $V_{c,3}$ in $layer(3)$ can be considered as to be derived from $V_{a,2}$, $V_{b,2}$ and $V_{c,2}$ during the construction of $layer(3)$ upon $(layer(2)$. The vertices of triangle $T_{l,3}$ emerge newly in the construction, and between them there is no straightforward correspondence with vertices in $layer(2)$. The connections between corresponding vertices are visualized in Fig. 5.14 as dotted lines; these connections will be indicated as the longitudinal elements. In Table 5.1 the list of vertex indices lists for Fig. 5.14 is displayed. The corresponding vertex indices have the same (horizontal) vertex index. The longitudinal elements can be reconstructed using this correspondence. The same table notation is applied for the list of triangle lists in Table 5.2. In this representation it is possible to find the neighbouring triangles of a given triangle in the multi-layered structure, and the (possibly existing) corresponding triangle in the succeeding and preceding layer can be detected without traversing all lists of triangles.

The longitudinal elements

In the case that $layer(1)$ was constructed upon $layer(2)$ and $layer(2)$ upon $layer(3)$, the list of vertex indices lists can be described as shown in Table 5.3.

The *null* signs in this table show that during the construction some of the series of longitudinal elements are interrupted, for example the series $V_{f,3}$, $V_{f,2}$ has no corresponding vertex on $layer(1)$.

Layered biological objects

The representation discussed in this section is now suitable to represent a model of a biological object, as described in Sect. 5.2. The sponge can be considered as built in layers, each layer consisting of edges (in the sponges the spicula) which are arranged in triangles. When a new layer emerges upon a preceding layer, it is possible to represent the subdivisions which are necessary to conserve a tessellation of nearly equilateral triangles. The construction of new layers and the subdivision of "non-fitting" triangles will be discussed further in the next section.

5.5 The Lattice Representation of a Volume Tessellated with Triangles

With the data structures discussed in the two previous sections it is possible to represent a geometric model of an organism that is constructed of layers of triangular tessellations. This representation in continuous 3D coordinates of the object has many advantages compared to a representation in discrete space. The growth process itself can be considered as a continuous process in time and the most natural way to model this process morphologically is by using a geometric model which allows continuous increase in size. Many geometric operations, for example transformations, can be done easily in continuous space. Some of these operations, for example rotation and scaling, become less trivial for objects represented in discrete space. However, for some purposes it is convenient to use a discrete version of the object, the lattice model, next to the geometric representation in continuous space.

The physical environment

In order to determine the influence of the physical environment on the growth process it is often necessary to use a discrete version of the model. In Sect. 3.7.2 it was demonstrated that the nutrient distribution around the object can be determined by solving the Laplace equation in discrete space. In many cases the influence of the physical environment can be described by partial differential equations like the Laplace equation in diffusion processes or the Navier-Stokes equation in a moving fluid. In many solution techniques for these equations, solutions are approximated in discrete space (see Ames 1977; Niemeyer et al. 1984; Doolen 1990).

In the Sects. 5.6.8 and 5.6.9 the influence of the physical environment will be modelled by using a lattice version of the geometric model. For this purpose the geometric model is mapped on a 3D lattice. The computations which can be done most conveniently in a 3D lattice are carried out. The results of these computations (local light intensities and nutrient concentrations) are used in a later stage again in the original model in continuous space.

Visualization of sections

A nice feature of the lattice representation of an object is that it offers the possibility to visualize sections made in an arbitrary plane. In an object that is represented in geometric space by a multi-layered triangular tessellation the lattice sites which belong to different layers can be marked. After this a section can be made through the object and the different

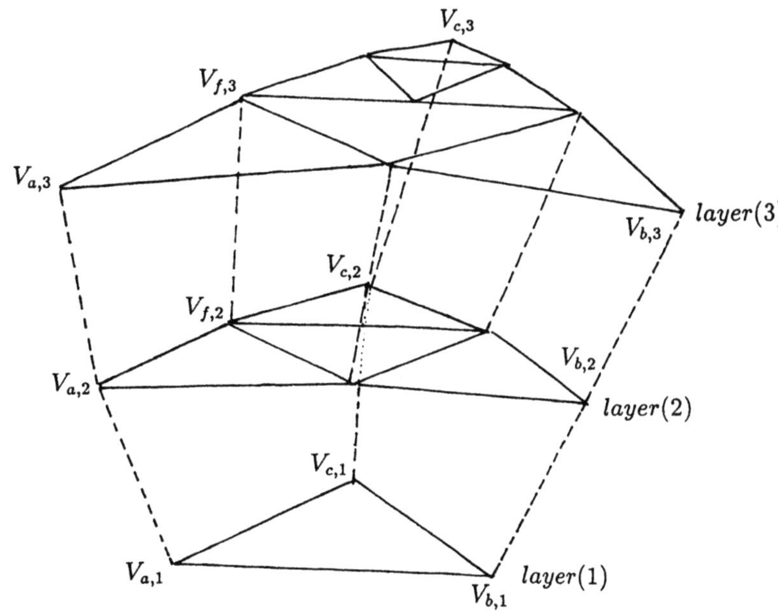

Fig. 5.14. An example of a layered system where a triangular tessellation is constructed upon another triangular tessellation

Table 5.1. List of vertex indices lists for the multi-layer structure in Fig. 5.14

layer index	1	2	3	4	5	6	7	8	9
1	$V_{a,1}$	$V_{b,1}$	$V_{c,1}$						
2	$V_{a,2}$	$V_{b,2}$	$V_{c,2}$	$V_{d,2}$	$V_{e,2}$	$V_{f,2}$			
3	$V_{a,3}$	$V_{b,3}$	$V_{c,3}$	$V_{d,3}$	$V_{e,3}$	$V_{f,3}$	$V_{g,3}$	$V_{h,3}$	$V_{l,3}$

Table 5.2. List of triangle lists for the multi-layer structure in Fig. 5.14

layer index	1	2	3	4	5	6	7
1	$T_{a,1}$						
2	$T_{b,2}$	$T_{c,2}$	$T_{d,2}$	$T_{e,2}$			
3	$T_{b,3}$	$T_{c,3}$	$T_{f,3}$	$T_{g,3}$	$T_{h,3}$	$T_{l,3}$	$T_{e,3}$

Table 5.3. List of vertex indices lists for the multi-layer structure in Fig. 5.14. In this table is assumed that $layer(1)$ is constructed upon $layer(2)$ and $layer(2)$ upon $layer(3)$.

layer index	1	2	3	4	5	6	7	8	9
3	$V_{a,3}$	$V_{b,3}$	$V_{c,3}$	$V_{d,3}$	$V_{e,3}$	$V_{f,3}$	$V_{g,3}$	$V_{h,3}$	$V_{l,3}$
2	$V_{a,2}$	$V_{b,2}$	$V_{c,2}$	$V_{d,2}$	$V_{e,2}$	$V_{f,2}$	*null*	*null*	*null*
1	$V_{a,1}$	$V_{b,1}$	$V_{c,1}$	*null*	*null*	*null*	*null*	*null*	*null*

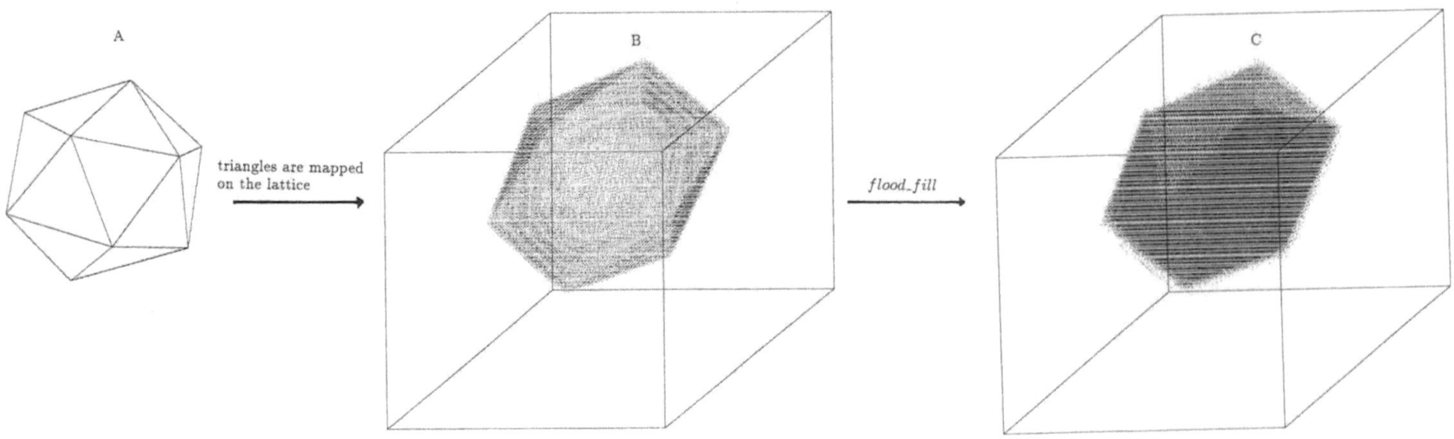

layers can be visualized. The simulated objects which are discussed in the following subsections can be sectioned with this method. It becomes possible to compare these sections with the actual sections as shown in Fig. 3.9 and Fig. 3.11B.

5.5.1 The Lattice Model

The discrete version of the surface

In the mapping of the geometric model on the 3D lattice, 3D versions of the Bresenham algorithm and scan-line filling algorithms are used. Algorithms which can be used for this purpose are described by Kaufman (1987, 1988). The mapping of the triangles of an icosahedron, represented in continuous space, on a 3D lattice of 100^3 sites is shown in Fig. 5.15. In the first step the triangles are drawn with a 3D scan-line filling algorithm in the lattice. After this step, the surface of the icosahedron is represented by occupied sites in the lattice, and the other sites in the lattice are in the state "unoccupied". In the next step the discrete version of the surface of the icosahedron is filled.

Filling the inside of the surface

The inside of the discrete version of the surface is filled using a procedure known in computer graphics literature as the flood-fill algorithm. The flood-fill starts with a lattice site inside the surface (the "seed", with lattice coordinates i_seed, j_seed, k_seed) and this site is set to the state "recently_occupied". The algorithm proceeds by testing the 6 direct neighbours of the sites which are in the state "recently_occupied". The algorithm stops as soon as no unoccupied neighbours are found anymore and the boolean $new\_added\_lattice\_sites$ remains $FALSE$. The algorithm is described in pseudo code below:

Fig. 5.15. Mapping of an icosahedron, represented in continuous space, on a 3D lattice with 100^3 sites. In A the geometric model is displayed, in B the triangles of the geometric model are represented in the lattice model, and in C the inside of the discrete version of the icosahedron is filled.

(5.17)

```
flood_fill( lattice with discrete version of the surface represented
by lattice sites in the state "occupied" ) {
  lattice[i_seed][j_seed][k_seed] = "recently_occupied";
  do {
    new_added_lattice_sites = FALSE;
    for each lattice site with coordinates (i,j,k) {
      if ( lattice[i][j][k] == "recently_occupied" ){
        if ( lattice[i-1][j][k] == "unoccupied" ){
          lattice[i-1][j][k] = "recently_occupied";
          new_added_lattice_sites = TRUE; }
        if ( lattice[i+1][j][k] == "unoccupied" ){
          lattice[i+1][j][k] = "recently_occupied";
          new_added_lattice_sites = TRUE; }
        if ( lattice[i][j-1][k] == "unoccupied" ){
          lattice[i][j-1][k] = "recently_occupied";
          new_added_lattice_sites = TRUE; }
        if ( lattice[i][j+1][k] == "unoccupied" ){
          lattice[i-1][j+1][k] = "recently_occupied";
          new_added_lattice_sites = TRUE; }
        if ( lattice[i][j][k-1] == "unoccupied" ){
          lattice[i][j][k-1] = "recently_occupied";
          new_added_lattice_sites = TRUE; }
        if ( lattice[i][j][k+1] == "unoccupied" ){
          lattice[i][j][k+1] = "recently_occupied";
          new_added_lattice_sites = TRUE; }
        lattice[i][j][k] = "occupied";
      }
    }
  while ( new_added_lattice_sites );
  all lattice sites in the state "recently_occupied" are changed
  into the state "occupied";
} end flood_fill
```

Preconditions for the flood-fill procedure

The result of the flood-fill operation is shown in the picture at the right in Fig. 5.15. The flood-fill procedure only works correctly when the discrete version of the surface in the lattice satisfies the following two preconditions: the surface should be a manifold and it is assumed that the scan-filling algorithm in the first stage in Fig. 5.15 worked correctly and did not leave "6-connective leaks" in the discrete versions of the triangles.

The resolution which is necessary in the lattice model depends on the types of computations in which the lattice model is used. For the storage of a lattice with 512^3 sites and an 8-bit representation of each lattice site it is necessary to allocate 128 Mbytes. For practical reasons such a huge allocation will often not be possible on the available hardware. An additional problem is that in many cases it will be necessary to maintain two copies of the lattice model during the computations: one copy with the old states and a copy with the updated versions of the lattice sites (see also Sect. 2.4 on solving the Laplace equation; for this problem two lattices are used).

Storage of the lattice

5.5.2 The Virtual Lattice, a Subdivision of Space

The hardware limitations mentioned in the previous section can be overcome by using the property that the considered objects often only occupy a small part of the lattice and the whole structure can be represented in a more compressed way. In an alternative representation a lattice is used with $lattice\_size^3$ lattice units, each unit can be further subdivided into 8 subunits, and each subunit can in turn again be subdivided into 8 smaller subunits. The value of $lattice\_size$ is chosen in such a way that the lattice can easily be allocated on the available hardware. Each unit or subunit can be in any of three states: "unoccupied", "occupied", or "subdivided". In the last case the unit or subunit contains an index which refers to a subunit in a list containing the state of the subunit. The subunit in the list can be again in the three possible states.

Subunits

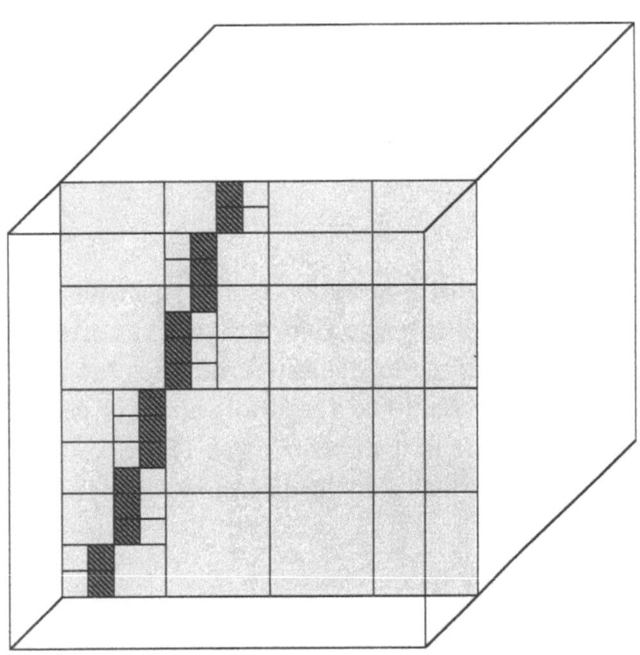

Fig. 5.16. An edge, visualized with a resolution of 16^3 lattice sites, represented with a lattice consisting of 4^3 sites and an additional list of 12 subunits

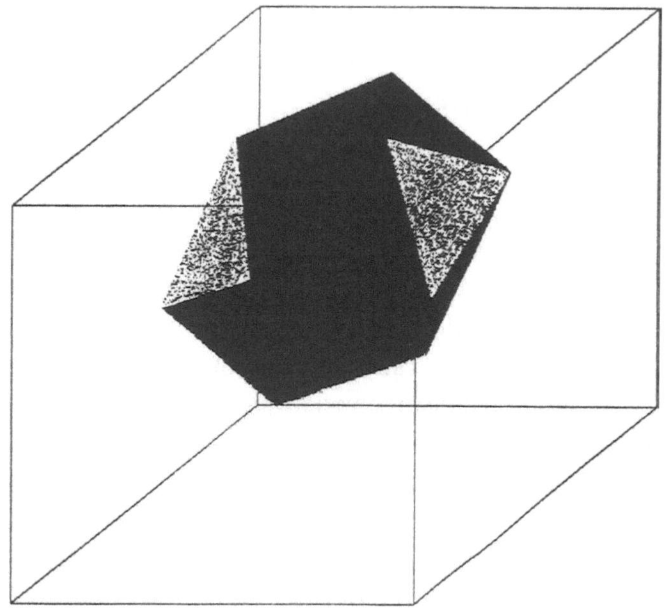

Fig. 5.17. The icosahedron volume from Fig. 5.15C visualized with a resolution of 400^3 lattice sites and represented with a lattice consisting of 100^3 sites and an additional list of 106 081 subunits

In Fig. 5.16 an example is shown how an edge, visualized with a resolution of 16^3 lattice sites, can be represented with a lattice of 4^3 sites and a list of 12 subunits. Only the sites near or at the edge occur in the list. The length of the list depends, of course, on the complexity of the object represented in the lattice. In the objects which will be shown in later sections, this method yields an enormous saving in memory. It also becomes possible to develop special versions of algorithms, for example the flood-fill algorithm shown in the previous section, which can traverse the 3D lattice relatively fast. In Fig 5.17 the volume of Fig. 5.15C is visualized in a virtual lattice with 400^3 sites. The surface was represented by a lattice of 100^3 sites and an additional list with subunits of 106 081 elements. In this additional list only the sites from the icosahedron volume are represented which are at or near the surface. Most of the inside of the volume is represented by units of the 100^3 lattice which are in the state "occupied", while most of the outside of the volume is represented as "unoccupied" sites in the 100^3 lattice. Only where a higher resolution is necessary, at and near the icosahedron surface, are sites from the 100^3 lattice in the state "subdivided" and refer to subunits from the additional list. In a version of the flood-fill algorithm which uses this virtual matrix, the step size when the lattice is traversed can be adapted. Outside and inside the icosahedron surface the step size will be a lattice unit of the 100^3 lattice, while near and at the surface the step sized is decreased to that of 400^3 lattice.

Flood-fill algorithms

5.6 An Iterative Geometric Construction Simulating the Radiate Accretive Growth Process of a Branching Organism

In this section the development of a 3D model for radiate accretive growth is discussed (see also Kaandorp 1993b). In this model the representation method for multi-layered triangular tessellations, discussed in the two previous sections, will be used to represent the objects in the iteration process. In the model a skeleton structure, as found in *Haliclona simulans*, is simulated. In the model a layer, consisting of a triangular tessellation, is constructed upon a preceding layer. In this simulation model the tangential spicula of *Haliclona simulans* are represented as tangential edges with an approximately constant size s (5.9), while the longitudinal spicula bundles are mimicked as edges with a variable length. The multi-layered structure (compare Fig. 5.14) is an imitation of the structure of *Haliclona simulans*, which may be transformed (as explained in Sect. 5.2) into a multi-layered structure of pentagons and hexagons. This structure can serve as a more general model for organisms with radiate accretive growth. The lattice representation of a triangulated object, as discussed in Sect. 5.5, will be used to determine the influence of the physical environment (light and nutrient distribution) on the growth process.

5.6.1 The Initiator

As an initiator of the iterative geometric constructions shown in this section, a triangulated sphere is used. The triangulated sphere can be derived from the icosahedron. In the icosahedron all vertices are situated on the surface of a sphere. When the triangles of the icosahedron are subdivided into four new triangles (compare the subdivision shown in Fig. 5.10) and the resulting new vertices are projected on the sphere, exactly enclosing the original icosahedron, a triangulated sphere is obtained. This procedure can be repeated leading to a series of spheres (see Fig. 5.18). The original icosahedron is shown at the left side of the picture. The series of triangulated spheres is also known as a series of icosahedral geodesic domes with different frequencies (see Wenninger 1979). The spheres are tiled with approximately equilateral triangles. The triangles are not completely equilateral; they can be organized in pentagons and hexagons. A convex object cannot be tiled with either pentagons or hexagons only (see also Wenninger 1971, Lord and Wilson 1984). The incenters of the pentagonal and hexagonal groups of triangles are called the pentavalent and hexava-

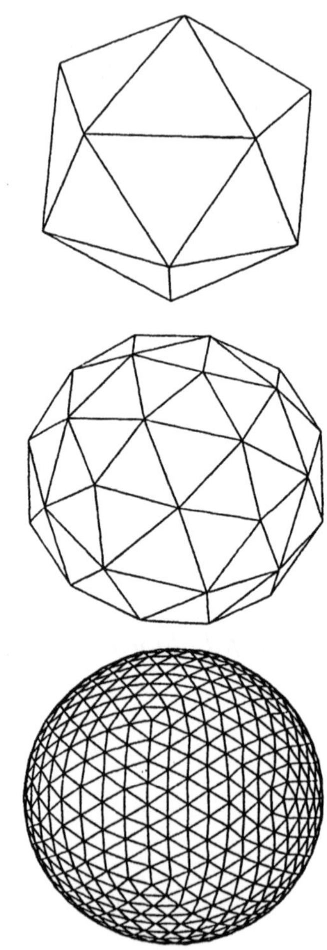

Fig. 5.18. A series of triangulated spheres derived from an icosahedron (displayed at the top of the figure) by subdivision of a triangle of a preceding sphere into four new ones. The subsequent subdivisions of the icosahedron serve as initial polyhedrons for the iteration process

Frequency of a triangulated sphere

lent vertices of the triangulated spheres. The frequency of such a sphere can be determined by counting the number of edges, using the shortest path, between two neighbouring pentavalent vertices. The frequency of the spheres depicted in Fig. 5.18 is from left to the right respectively 1, 2, and 8. The pentavalent vertices in this figure are the original vertices of the icosahedron.

Aulonia

Members of this series of icosahedral domes are often found in nature. An example is the spherical radiolarian *Aulonia* shown in Fig. 2.7. When the triangles of the sphere are organized in pentagons and hexagons, a structure is obtained closely resembling *Aulonia*. Another famous example of spheres, tiled with pentagons and hexagons, which can be derived from the series of icosahedral domes for given frequencies, are the fullerenes (see Carl and Smalley 1991). In these C-molecules, from which the "buckyball" is the most famous example, the C-atoms are organized in a sphere tiled with pentagons and hexagons.

Protrusions in an encrusting layer

The biological relevance of the spherical initiator for the object with radiate accretive growth is derived from the observation that in many of these organisms the growth process starts from hemispherical protrusions, which develop in an encrusting layer. The arrangement of the tangential elements in the triangulated spheres corresponds with the penta-hexagonal arrangements, which can be observed in tangential views of organisms with radiate accretive growth (see Figs. 3.10B, 3.13A, and 5.8).

The 3D model for radiate accretive growth is developed stepwise, applying a comparable strategy as in the development of the 2D model (see Fig. 3.17); this development is shown in Fig. 5.19. The initiators used in all constructions displayed in this section are all n-frequency derivates from the the icosahedron.

5.6.2 The Basic Construction: the Generator

The basic construction is depicted in Fig. 5.20. In this picture a new longitudinal element, with length l, is constructed upon a preceding layer. The longitudinal element is constructed perpendicular to the preceding layer by determining the mean value of the direction of the normal vectors in the collection $set\_triangles(T_{i,j}, V_{i,j})$: the set of triangles surrounding the vertex $V_{i,j}$ (5.14). The length l is determined by the *generator processing function* from (3.1). The angle α is used as an argument in the *generator processing function* ($local\_inf$ in Fig. 2.40), the angle between the longitudinal element and the vertical. The construction generates a new vertex $V_{i,j+1}$; the newly generated vertices together define new triangles $T_{i,j+1}$. The construction is basically similar to the one shown in Fig. 3.19.

A difference is that in Fig. 3.19 a new "fertile" tangential edge and a "non-fertile" longitudinal one are added to the originally "fertile" tangential edges, while in Fig. 5.20 originally "fertile" vertices generate new "fertile" vertices. The new vertices define a new longitudinal element and new tangential ones. The tangential elements are by definition connected which makes the *continuity rule* as applied in in the 2D construction (see Sect. 3.6.2) superfluous. The construction is described in the following replacement system:

Fig. 5.19. Diagram showing the development of 3D growth models for organisms with radiate accretive growth

$$
\begin{aligned}
initiator &= \mathcal{V} = \{(V_{1,1}, F); \cdots (V_{42,1}, F); \} \qquad (5.18)\\
&\quad \mathcal{T} = \{T_{1,1}; \cdots T_{80,1}; \}\\
generator &= (V_{i,j}, F); \rightarrow (V_{i,j}, NF); (V_{i,j+1}, F);\\
&\quad (V_{a,j}, F); \rightarrow (V_{a,j}, NF); (V_{a,j+1}, F);\\
&\quad (V_{b,j}, F); \rightarrow (V_{b,j}, NF); (V_{b,j+1}, F);\\
&\quad T_{i,j} = ((V_{i,j}, V_{a,j}, V_{b,j}), \cdots)\\
&\quad T_{i,j+1} = ((V_{i,j+1}, V_{a,j+1}, V_{b,j+1}), \cdots)\\
&\quad T_{i,j}; \rightarrow T_{i,j}; T_{i,j+1};
\end{aligned}
$$

In this system the construction starts with a triangulated sphere consisting of 80 faces. First the new vertices ($V_{i,j+1}$, $V_{a,j+1}$, $V_{b,j+1}$) are generated, and after this step a new triangle $T_{i,j+1}$ can be constructed between the three neighbouring new vertices.

The result of the construction, after a few iteration steps, is displayed in Fig. 5.19A. The biological interpretation of the *generator processing function* is explained in the Sect. 3.6.1, on the basic construction applied in the 2D model.

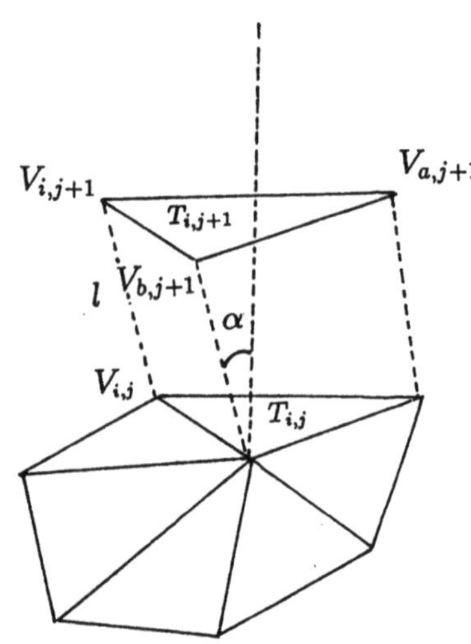

Fig. 5.20. The basic construction applied in the 3D models for radiate accretive growth. In the construction a new vertex $V_{i,j+1}$ is constructed upon a vertex $V_{i,j}$ from the preceding layer: the construction defines a new longitudinal element (length l) and a new triangle $T_{i,j+1}$, consisting of tangential elements.

5.6.3 Isotropic Growth and the Insertion of New Elements

In the object in Fig. 5.19A it can be seen that the tangential elements dilate further with each iteration step. To obtain an object where the surface is tessellated with triangles with nearly equal-sized tangential elements an additional rule is necessary. In the simplest case this problem can be solved by subdividing any triangle where all three edges exceed the maximum value 1.5s allowed for a tangential element (5.9) into four triangles. Examples of such subdivisions are shown in Figs. 5.10 and 5.14. The effect of setting a maximum size to a tangential element is depicted in Fig. 5.19B. The subdivision of excessively large triangles can be compared with the *insertion rule* in Sect. 3.6.2 used to preserve

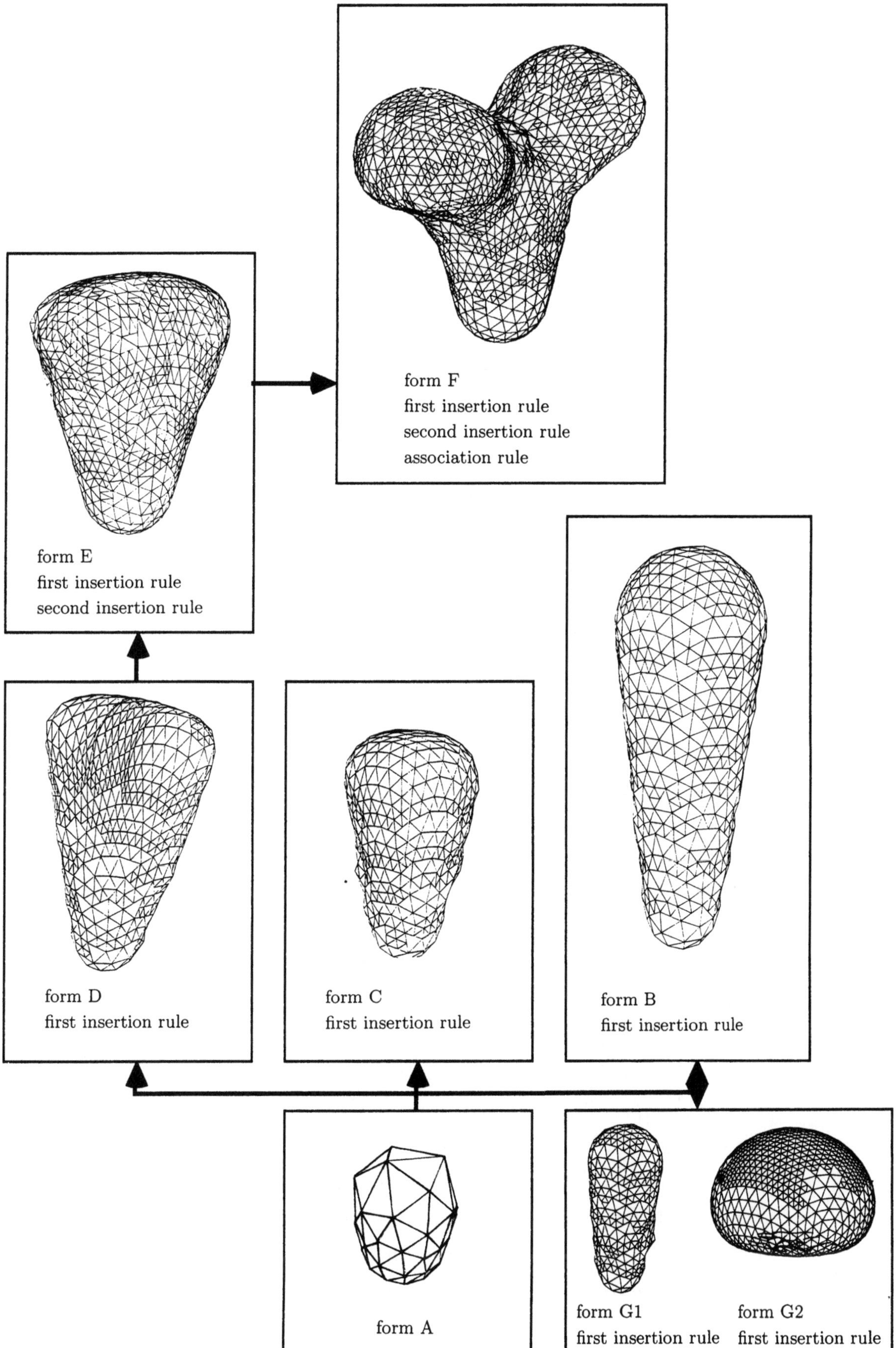
form F
first insertion rule
second insertion rule
association rule
form E
first insertion rule
second insertion rule
form D
first insertion rule
form C
first insertion rule
form B
first insertion rule
form A
form G1
first insertion rule
form G2
first insertion rule

coherence in the 2D skeleton. The insertion of new triangles and the inclusion of a threshold *inhibition_level* for the longitudinal element (this threshold will be discussed below) is described in the following replacement system:

$$
\begin{aligned}
\mathit{initiator} &= \mathcal{V} = \{(V_{1,1}, F); \cdots (V_{42,1}, F); \} \qquad (5.19)\\
&\quad \mathcal{T} = \{T_{1,1}; \cdots T_{80,1}; \}\\
\mathit{generator} &= (V_{a,j}, F); \rightarrow\\
&\quad \text{if } (l > \mathit{inhibition\_level}) \text{ then } (V_{a,j}, NF); (V_{a,j+1}, F);\\
&\quad \text{else } (V_{a,j}, NF); (V_{a,j+1}, SF);\\
&\quad (V_{b,j}, F); \rightarrow\\
&\quad \text{if } (l > \mathit{inhibition\_level}) \text{ then } (V_{b,j}, NF); V_{b,j+1}, F);\\
&\quad \text{else } (V_{b,j}, NF); (V_{b,j+1}, SF);\\
&\quad (V_{c,j}, F); \rightarrow\\
&\quad \text{if } (l > \mathit{inhibition\_level}) \text{ then } (V_{c,j}, NF); (c_{b,j+1}, F);\\
&\quad \text{else } (V_{c,j}, NF); (V_{c,j+1}, SF);\\
&\quad (V_{k1,j}, NF); \rightarrow (V_{k1,j}, NF);\\
&\quad (V_{k2,j}, SF); \rightarrow (V_{k2,j}, SF);\\
&\quad T_{i,j} = ((V_{a,j}, V_{b,j}, V_{b,j}), \cdots)\\
&\quad T_{i,j+1} = ((V_{a,j+1}, V_{b,j+1}, V_{c,j+1}), \cdots)\\
&\quad T_{i,j}; \rightarrow T_{i,j}; T_{i,j+1};\\
\mathit{3D\ insertion} &\\
\mathit{rule}\ 1 &= \text{if } ((\| V_{a,j+1}, V_{b,j+1} \| > 1.5s)\ \&\&\\
&\quad (\| V_{b,j+1}, V_{c,j+1} \| > 1.5s)\ \&\&\\
&\quad (\| V_{c,j+1}, V_{a,j+1} \| > 1.5s)) \text{ then}\\
&\quad (V_{a,j+1}, F); (V_{b,j+1}, F); (V_{c,j+1}, F); \rightarrow\\
&\quad (V_{a,j+1}, F); (V_{na,j+1}, F); (V_{b,j+1}, F);\\
&\quad (V_{nb,j+1}, F); (V_{c,j+1}, F); (V_{nc,j+1}, F);\\
&\quad T1 = ((V_{a,j+1}, V_{na,j+1}, V_{nc,j+1}, (\cdots), (T5, T7, 0))\\
&\quad T2 = ((V_{na,j+1}, V_{b,j+1}, V_{nb,j+1}), (\cdots), (T5, T6, 0))\\
&\quad T3 = ((V_{nc,j+1}, V_{nb,j+1}, V_{c,j+1}), (\cdots), (T6, T7, 0))\\
&\quad T4 = ((V_{na,j+1}, V_{nb,j+1}, V_{nc,j+1}), (\cdots), (\mathit{null}, \mathit{null}, 0))\\
&\quad T5 = ((T1, T2, \mathit{null}), (\cdots), (\mathit{null}, \mathit{null}, 1))\\
&\quad T6 = ((\mathit{null}, T2, T3), (\cdots), (\mathit{null}, \mathit{null}, 1))\\
&\quad T7 = ((T1, \mathit{null}, T3), (\cdots), (\mathit{null}, \mathit{null}, 1))\\
&\quad T_{i,j+1}; \rightarrow T1; \cdots T7;
\end{aligned}
$$

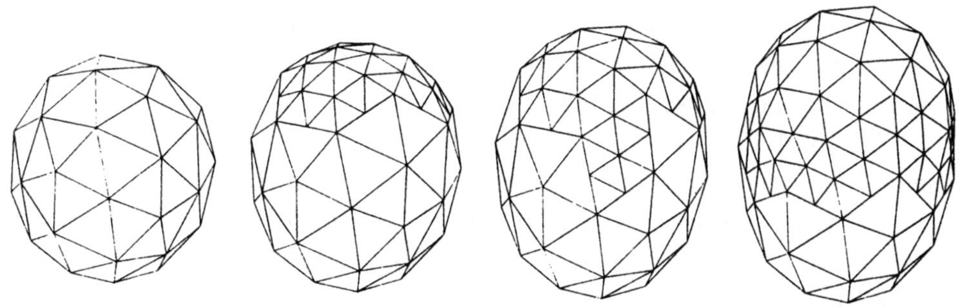

Fig. 5.21. Objects emerging subsequently in the iteration process by applying model Fig. 5.19

3D insertion rule

The consequence of this rule (*3D insertion rule 1*) is that new triangles and longitudinal elements are generated. In Fig. 5.21 a few column-like objects are shown which emerge in subsequent iteration steps. Each time a new vertex is generated in the iteration process, the starting point of a corresponding new longitudinal element is defined on the growth layer.

The biological meaning of this *3D-insertion rule* is that in reality the surface of the organism is tessellated with discrete skeleton elements, which can vary only somewhat in size (see Figs. 3.10B, 3.13A, and 5.8). The choice of the lower and upper limit in (5.9) was an arbitrary one; in reality these limits are typically species-specific properties.

When the *generator processing function* (3.1) is applied a layered structure is generated, where the distance between the subsequent layers can become zero at the sides of a column-shaped object. The length of the longitudinal element l will become zero at a vertex $V_{k,j}$, where the mean value of the directions of the normal vectors of the the triangles surrounding the vertex ($set\_triangles(V_{k,j}, T_{i,j})$ in (5.14)) is perpendicular to the direction of the y-axis.

Threshold value of the longitudinal element

In order to avoid the generation of a multi-layered structure with arbitrarily short longitudinal elements, a *generator-processing functior* with a threshold value ($inhibition\_level$ in (3.9)) can be used. In analogy with the 2D model (see Sect. 3.6.3) the status of the fertile vertices is set to the state "non-active". The surface of the object is limited by a mesh with fertile vertices in either the state "active" (F in (5.19)) or "non-active" (SF in (5.19)). These "non-active" fertile vertices can be used in a more advanced version of the 3D model, where the "non-active" vertices are set to state "active" again and can participate in a secondary growth process (see also Sect. 3.6.3). The consequence of the introduction of "non-active" fertile vertices is that in some parts of the multi-layered structure the subsequent layers may coincide in the areas with "non-active" vertices. The result of applying a *generator-processing function* with a threshold is an object as shown in Fig. 5.19B. The difference between objects generated with a *generator-processing function* with and without this threshold can only be observed when longitudinal sections of both objects are compared (see Figs. 3.17B and 3.17C), where in these sections the thickness of subsequent layers is visible.

The biological interpretation of the threshold rule is discussed in Sect. 3.6.3: in organisms with radiate accretive growth, where the skeleton elements are secreted internally (as in the *Haliclona* species), the longitudinal fibres cannot become too short, since these fibres are built from discrete skeleton elements.

When a longitudinal section is made through the object depicted in Fig. 5.19B, parallel to the axis of growth (in this case the vertical), the layered structure becomes visible. In Fig. 5.22 a part of the object is removed to show this layered structure. The image obtained from this longitudinal section corresponds with the 2D simulations from Chap. 3. When another longitudinal section is made through the object in Fig. 5.22 and the section is made parallel to the axis of growth and includes the centre of the column, the same image will be obtained. The type of *generator processing function* (3.1) used in this object, where different longitudinal sections remain equal, can be defined as "isotropic". In the actual objects a characteristic of isotropic growth is that the growth function empirically derived from longitudinal sections (see Fig. 3.20) turns out to be the same in different section planes.

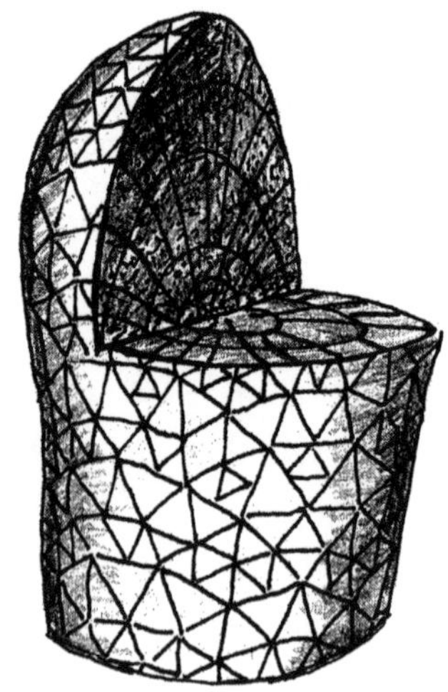

Fig. 5.22. Longitudinal section through a column-shaped object (see Fig. 5.19B) where a part of the object is removed to show the layered structure

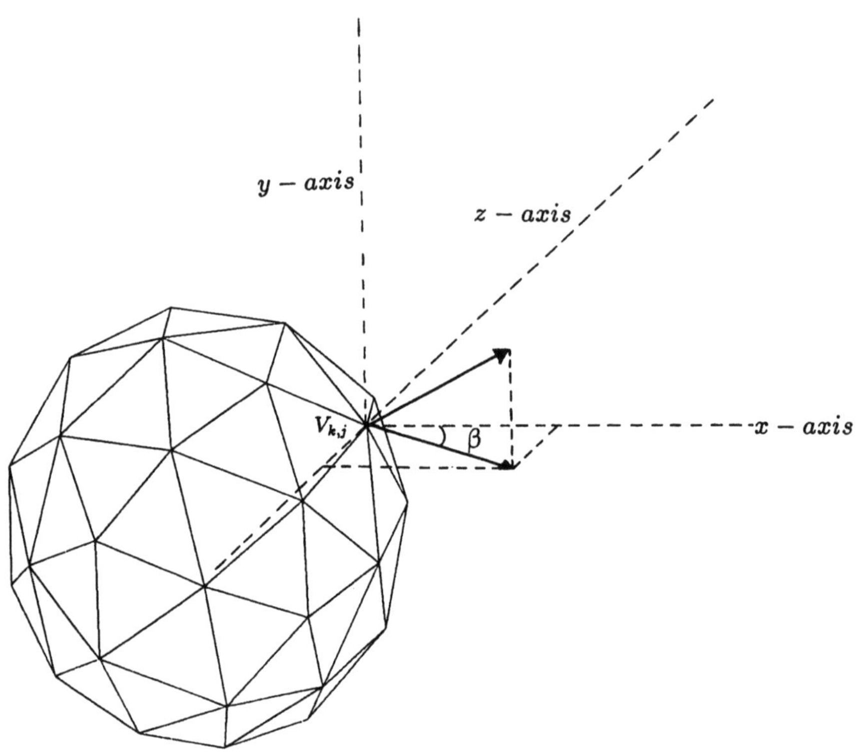

Fig. 5.23. Determination of the angle β between the projection of the mean normal vector on the xz plane, in a vertex $V_{k,j}$ and the horizontal (indicated as the x-axis). In the model the direction of the flow corresponds to the z-axis.

5.6.4 Anisotropic Growth and the Insertion of New Elements

Club-shaped objects

The Model. In Fig. 5.19B the curvature of the top of the object remains unchanged in the iteration process. In order to obtain objects which flatten slightly at the top, a *generator-processing function* like the one from (3.11) can be used. The result is a club-shaped or clavate object, as shown in Fig. 5.19C. In this object the degree of widening can be controlled with the parameter w. In this clavate object a disc of equal-sized longitudinal elements appears instead of one maximum value. A longitudinal section, made parallel to the axis of growth and including the centre of the object, will yield the same image in different planes. The function from (3.11) is an isotropic one.

An example of an anisotropic *generator-processing function* is shown in (5.20). In this function the object does not widen equally in all directions.

$$
\begin{aligned}
d &= w/\cos(\beta) \text{ for } 0 \le \beta \le \pi/2 \qquad (5.20)\\
f(\alpha, \beta) &= \begin{cases} 1.0 \text{ for } 0 \le \alpha \le (\pi/d) \\ \cos((\pi/2)/(\pi/2 - \pi/d) \cdot (\alpha - \pi/d)) \\ \text{for } (\pi/2 + \pi/d) < \alpha \le \pi \end{cases}\\
l &= \begin{cases} s \cdot f(\alpha, \beta) \text{ for } f(\alpha, \beta) > inhibition\_level \\ 0.0 \text{ for } f(\alpha, \beta) \le inhibition\_level \end{cases}\\
&\quad w > 2
\end{aligned}
$$

Widening effect

In this function the degree of widening d depends on the angle β (see Fig. 5.23) between projection of the mean normal vector in a vertex $V_{k,j}$ on the xz-plane and the horizontal (indicated as the x-axis). The angle β represents the angle between a longitudinal element and the flow direction, which corresponds with the direction of the z-axis. When $\beta = 0$ the widening effect, controlled by the parameter w, is maximal. In this case the function $f(\alpha, \beta)$ corresponds with the *generator-processing function* from (3.11). For $\beta = \pi/2$ the widening effect is minimal, d is infinitely large, and $f(\alpha, \beta)$ corresponds with the function $cos(\alpha)$. In Fig. 5.24 the shape of function (5.20) is displayed for various values of β. The result of applying this *generator-processing function* is a flattened clavate or planar object (see Fig. 5.19D). In the flattened object a pretzel-shaped area emerges of equal-sized longitudinal elements.

A side-effect which occurs in the generation of a flattened object (Fig. 5.19D) is that the triangles on the surface are not enlarged equally in all directions, as in the clavate object (Fig. 5.19C). The result is an object in which one or two edges of a triangle exceed $1.5s$ in (5.9) so that when

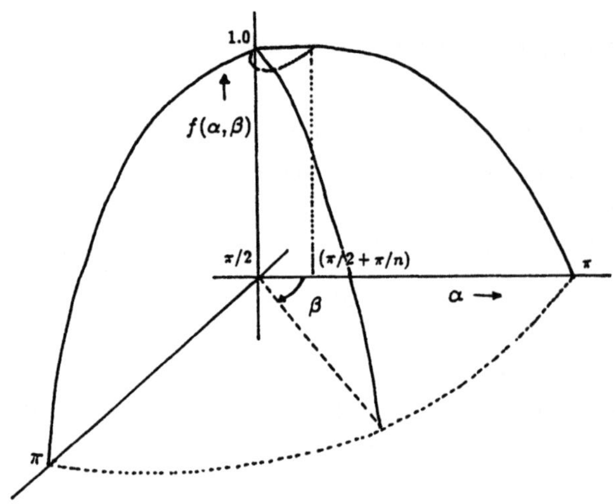

Fig. 5.24. Shape of the function $f(\alpha, \beta)$ (5.20)

the construction is continued a tessellation emerges where the triangles are far from approximately equilateral. The problem cannot simply be solved by the subdivision of the triangle into 4 smaller daughter triangles (*3D insertion rule 1*), for this operation would yield triangles with one or more too small (smaller than $0.5s$) edges. A solution is depicted in Fig. 5.25. Here two triangles, both with two edges exceeding $1.5s$ in (5.9), are subdivided into four daughter triangles. This subdivision method (*3D insertion rule 2*) does not disturb the penta-hexagonal organization of the tessellation by introducing for example an octavalent vertex, and all vertices remain quadri-, penta- or hexavalent.

Second 3D insertion rule

The Biological Objects. In Sect. 3.6.4 the biological relevance of the widening effect is discussed. The relation with the actual objects can be observed in Fig. 3.11B, where the branch widens before branching. Without widening, the organism would split up into ever smaller branches. In the longitudinal section it can be seen that the organism widens slightly, because there is a small area of equal longitudinal elements at the tip of the branch. The widening effect can work equally in all directions causing a club-shaped organism, which has some special implications for the branching process. The biological relation of this clavate form in the formation of branches will be discussed later on.

In most cases, in for example both *Haliclona* species, it can be observed that branches do not widen equally in all directions. Especially in *Haliclona oculata* (see Fig. 3.11B) the sponge widens more strongly in one plane than in the other, causing a flattened growth form. For many marine organisms the growth process takes place in an environment with tidal flows. The flow pattern for a laminar flow might occur as depicted in Fig. 5.26 (see Wainwright and Koehl 1976; Koehl 1976 and 1982; Vo-

Flattened growth forms

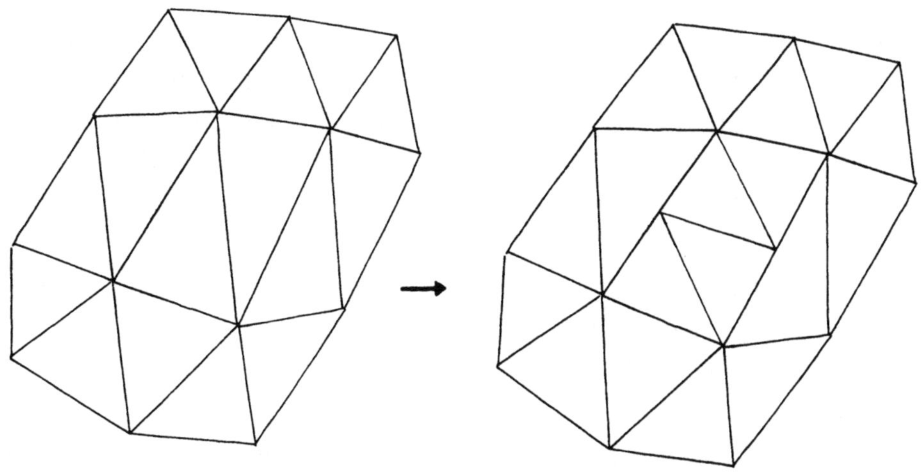

Fig. 5.25. Subdivision of two triangles, both with two edges exceeding $1.5s$ in (5.9), into four daughter triangles

gel 1983). The flow pattern reverses in direction, basically twice a day, because of the tidal movement. The flow velocities around the object will be zero at point A (flow perpendicular to the object) and increase towards the sides, and have a maximum at point B. The flow velocities are highest at the lateral sides of the organism, parallel to the flow, together with the supply of suspended material. This results in a relatively faster growth towards the lateral sides parallel to the flow. The growth velocity at a point in the organism depends on the angle β: the angle between the projection of the mean normal vector in a point on the xz plane and the x-axis (when the organism's growth axis is the y-axis and the flow is in the z-direction). In $f(\alpha, \beta)$ (5.20) the secretion of elements also depends on the angle β with the flow direction. In many marine sessile organisms the phenomenon can be observed that flattened growth forms emerge (see also Fig. 3.8), when the organism is positioned perpendicular to the prevailing flow direction (see Jackson 1979). The emergence of such a flattened form can be explained by assuming, as above, that the growth velocities are not equal in all directions, but depend on the position with respect to this flow direction.

The angle β

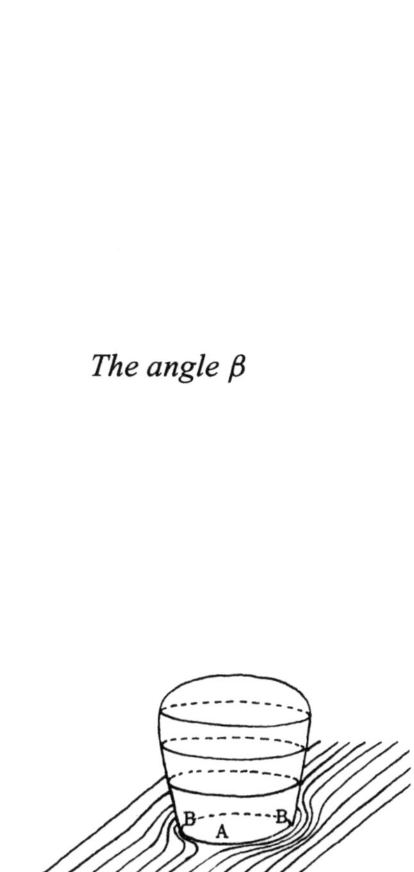

Fig. 5.26. Flow pattern around an organism growing in a laminar flow. The flow velocities around the object are zero at point A (flow perpendicular to the object), increase towards the sides, and have a maximum in point B.

The biological interpretation of the second *3D insertion rule* corresponds with the one for the first *3D insertion rule*. The preservation of the penta- hexagonal organization is based on observations made in tangential views of the biological objects (see Figs. 3.10B and 3.13A), where predominantly quadri-, penta- and hexavalent vertices are observed.

5.6.5 Formation of Branches

Branches are formed in analogy with the 2D model. For this purpose a series of estimations is made of the local radius of curvature in a vertex $V_{k,j}$. The first step is to collect a set *circle* of vertices surrounding the vertices in the set $set\_vertices(T_{i,j}, V_{k,j})$ (5.8). After this the set $set\_vertices(T_{i,j}, V_{k,j})$ is deleted from *circle*, a set of vertices is left which are situated approximately in a circle around the vertex $V_{k,j}$ (see Fig. 5.27). The algorithm in which the set *circle* is determined is described below:

collect_circle(circle, $T_{i,j}$, $V_{k,j}$ *)* { (5.21)
  circle = $set\_vertices(T_{i,j}, V_{k,j})$;
  st = $set\_triangles(T_{i,j}, V_{k,j})$;
  for each vertex $V_{k2,j} \in set\_vertices(T_{i,j}, V_{k,j})$ {
    take a triangle $T_{i2,j}$ from *st* in which the vertex $V_{k2,j}$ occurs;
    circle += $set\_vertices(T_{i2,j}, V_{k2,j})$;
  }
  circle -= $set\_vertices(T_{i,j}, V_{k,j})$;
} end *collect_circle*

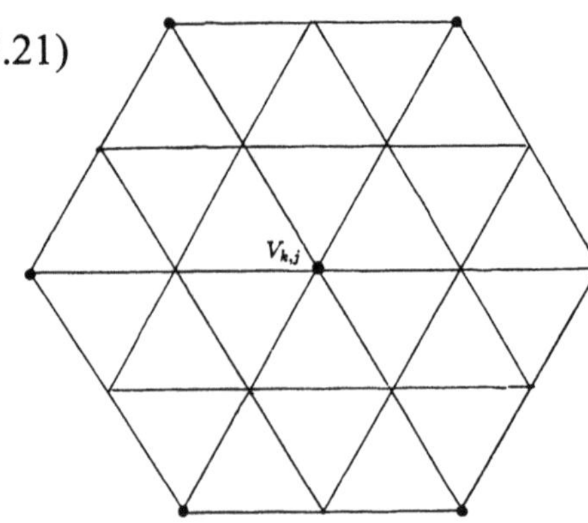

Fig. 5.27. Determination of a set of vertices, situated approximately in a circle around the vertex $V_{k,j}$. From these vertices 3 pairs (indicated with black dots) of vertices are selected, which are situated at a maximum distance from each other. The pairs and $V_{k,j}$ itself are used for the estimations of the radius of curvature.

From these vertices 3 pairs (indicated with black dots) of vertices are selected, which are situated at a maximum distance from each other. The vertices, can be considered as situated approximately on a circle. The vertices in a pair are situated approximately on a diameter of this circle. In the second step, for each pair and the vertex $V_{k,j}$ an estimation is made of the radius of curvature. This radius of curvature is estimated in a plane perpendicular to the mean normal vector in vertex $V_{k,j}$. For this purpose the pair of vertices is projected onto this plane. The estimation is done by constructing a circle through the two projected vertices and the vertex $V_{k,j}$. The radius of curvature is infinitely large when the projected pair of points and $V_{k,j}$ are situated on a line and is negative when they are on a concave surface. In the third step the radii of curvature are normalized using function $h(rad\_curv)$ of (3.14). In step four the measurements are summarized in one index ($curv\_index$) expressing the amount of contact with the environment. This index is formed by the product of the lowest value (low_norm_curv) and the average value (av_norm_curv) of the normalized radii of curvature:

$$curv\_index = low\_norm\_curv \cdot av\_norm\_curv \qquad (5.22)$$

Amount of contact with the environment

In the actual organism the amount of contact with the environment will decrease when the average radius of curvature increases. When a certain critical value (max_curv) in (3.14) is exceeded, which is related to the capabilities of the transport system, no nutrient at all will be transported and the growth velocity will become zero.

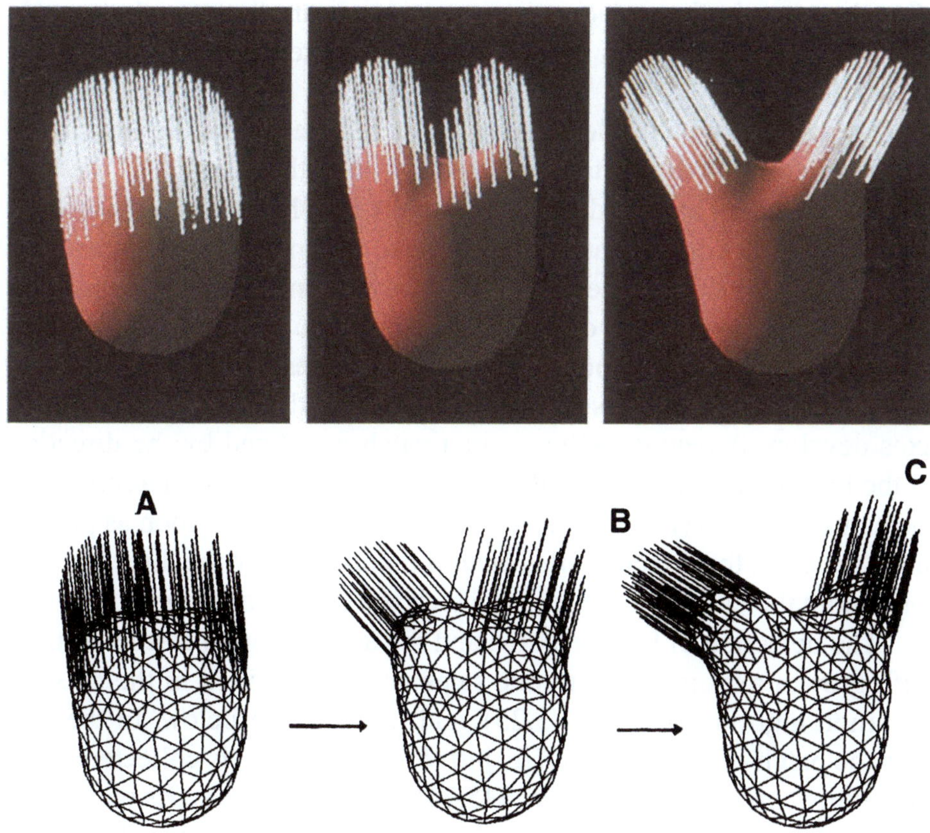

Fig. 5.28. Replacement of an old growth axis A by two new ones B and C. The fertile vertices in this picture are visualized with edges which represent the direction of the associated growth axis.

In Fig. 5.19F this estimation of the local radius of curvature is used to determine the length l of new of new longitudinal elements, which is now determined by a combination of $f(\alpha, \beta)$ and $curv\_index$:

$$l = \begin{cases} s \cdot f(\alpha, \beta) \cdot curv\_index \\ \text{for } f(\alpha, \beta) \cdot curv\_index > inhibition\_level \\ 0.0 \text{ for } f(\alpha, \beta) \cdot curv\_index \leq inhibition\_level \end{cases} \tag{5.23}$$

In this construction a flattened object is used, which widens more in the xy plane than in the xz plane. As soon as the radius of curvature exceeds the maximum allowed value, in the xy plane two local maxima are formed (see Fig. 5.28). In the 3D model the longitudinal elements with a local maximum value, or an area with equal-sized maximum values, are surrounded by smaller ones. In correspondence with the 2D model these local maxima are associated with new growth axes. In Fig. 5.28 the replacement of an old growth axis A by two new ones B and C is displayed. The fertile vertices in this figure are visualized with edges which represent the direction of the associated growth axis. The growth axis is an attribute

Replacement of growth axes

of each vertex in the replacement system (compare the *association rule* in the replacement system in (3.13)); this attribute is changed in the *3D association rule*.

3D association rule

The patches with fertile vertices in Fig. 5.28 represent the sites in the object where growth still takes place and where the length l of the longitudinal element has not dropped below the threshold $inhibition\_level$ in (5.23). After some iteration steps a local minimum develops and the patch of fertile vertices is separated into two new ones, because locally the maximum allowed radius of curvature max_curv in (3.14) is exceeded and the $curv\_index$ component in (5.22) becomes zero. In the two new patches the growth process continues independently and two new growth axes develop. A new growth axis in a patch is defined by the direction of the longitudinal element with a length l which is a local maximum. In the actual sections (Fig. 3.11B) it can be observed that as soon as protrusions develop, the direction of growth is determined by the direction of a longitudinal element with a length which is a local maximum.

The detection of local maxima and the association of fertile vertices with growth axes is done in three steps and is described in the axis association algorithms (5.24), (5.25), and (5.26) given below. In the first step the patch of vertices $fertile\_patch$ in the state F is collected in the procedure $collect\_fertile\_patch$ (5.24). The procedure starts with an arbitrary chosen vertex in the state F $V_{k1,j}$. In the algorithm two alternating lists $list\_a$ and $list\_b$ are used. In all three axis association algorithms, a vertex in combination with a triangle in which the vertex occurs is always used. As discussed in Sect. 5.3 this combination is necessary since the only possible way to access the individual vertices is to use the list $\mathcal{T}$: in this list the neighbour triangles and their vertices of a given triangle $T_{i,j}$ can be accessed, while the list $\mathcal{V}$ does not possess this feature. In (5.24) the algorithm proceeds by extending the patch of fertile vertices $fertile\_patch$ with new fertile vertices which surround the vertices at the borders of $fertile\_patch$. The new vertices and their corresponding triangles are collected in the sets sv and st respectively. The algorithm stops as soon no new fertile vertices are encountered and $fertile\_patch$ is bordered by vertices in the state SF.

Collection of a patch of fertile vertices

collect_fertile_patch($T_{i1,j}$, $V_{k1,j}$, *fertile_patch*){ (5.24)
  fertile_patch = ϕ;
  list_a, with triangle vertex pairs, is initialized with the pair ($T_{i1,j}$, $V_{k1,j}$);
  do {
    list_b, with triangle vertex pairs, is emptied;
    for each pair ($T_{i2,j}$, $V_{k2,j}$) in *list_a* {
      sv = *set_vertices*($T_{i2,j}$, $V_{k2,j}$);
      st = *set_triangles*($T_{i2,j}$, $V_{k2,j}$);
      sv -= *fertile_patch*;
      all vertices in the state *SF* are removed from *sv*;
      for each vertex $V_{k3,j}$ in *sv*{
        take a triangle $T_{i3,j}$ from *st* in which the vertex $V_{k3,j}$ occurs;
        add the pair ($T_{i3,j}$, $V_{k3,j}$) to *list_b*;
      }
      fertile_patch += *sv*;
      copy *list_b* to *list_a*;
    }
  } **while** (*list_b* is not empty);
} end *collect_fertile_patch*

Detection of local maxima

In the second step of the *3D association rule*, $fertile\_patch$ is processed by the procedure $find\_local\_maxima$ (5.25), in which the local maxima are detected in the patch. For this purpose the distance d between each vertex in $fertile\_patch$ in $layer(j)$ and its predecessor in $layer(j-1)$, the length of the longitudinal element, is measured. In the case the vertex $V_{k1,j}$ in $fertile\_patch$ was newly inserted in the first or second *3D insertion rule*, this corresponding predecessor does not exist. In the algorithm (5.25) one or two longitudinal elements, with indices $max1\_nr$ and $max2\_nr$ and with respective lengths $max1\_d$ and $max2\_d$ which are local maxima, are detected. As depicted in Fig. 5.24, the application of function $f(\alpha, \beta)$ may lead to an area of equal-sized local maximum values of lengths of longitudinal elements. In this case the potential maximum $max1\_nr$ is positioned in the centre of the area. When there is only one local maximum detected, as for example in the left object in Fig. 5.28, the vector $DA1$ with the direction of the maximum longitudinal element is determined. All vertices in the patch $fertile\_patch$ are associated with the growth axis $DA1$. In the case there are two local maxima detected (for example the right object in Fig. 5.28), two growth axes $DA1$ and $DA2$ are determined and $fertile\_patch$ is split into two new patches in the procedure $split\_patch$.

Association with growth axes

```
find_local_maxima(fertile_patch){                                    (5.25)
  max2_d = max1_d = 0;
  max2_nr = max1_nr = null;
  for each vertex V_{k1,j} ∈ fertile_patch {
    if ( vertex V_{k1,j-1} exists ){
      d = || V_{k1,j-1}, V_{k1,j} ||;
      if (d > max1_d){
        max2_d = max1_d;
        max2_nr = max1_nr;
        max1_d = d;
        max1_nr = k1;
        if ( longitudinal element (V_{k1,j-1}, V_{k1,j}) is in a
        patch with longitudinal elements with equal lengths ){
          the index max1_nr is set to the index of the
          longitudinal element situated centrally in the patch;
        }
      }
    }
  }

  if ( max_nr2 == null ){
    DA1 = V_{max_nr1,j} - V_{max_nr1,j-1};
    all vertices in fertile_patch are associated with the vector DA1;
  }
  else {
    take a triangle T_{i1,j} in which the vertex V_{max_nr1,j} occurs;
    DA1 = V_{max_nr1,j} - V_{max_nr1,j-1};
    take a triangle T_{i2,j} in which the vertex V_{max_nr2,j} occurs;
    DA2 = V_{max_nr2,j} - V_{max_nr2,j-1};
    split_patch( T_{i1,j}, V_{max_nr1,j}, DA1, T_{i2,j}, V_{max_nr2,j}, DA2 );
  }
} end find_local_maxima
```

The procedure *split_patch* works in a similar way as *collect_fertile_patch* and is also based on set operations. The procedure is initialized with the two vertices that are local maxima and both vertices are associated with their corresponding growth axes $DA1$ and $DA2$. The procedure involves collecting new vertices in the state F in the set sv around a vertex $V_{k1,j}$ from the collection of fertile vertices. The vertices in sv are associated with the growth axis $DAXIS$, which is an attribute of $V_{k1,j}$. The algorithm stops as soon no new vertices in the state F are found and the investigated area of fertile vertices $patch$ is bordered by vertices in the state SF.

Axis attribute

split_patch($T_{i1,j}$, $V_{max\_nr1,j}$, *DA1*, $T_{i2,j}$, $V_{max\_nr2,j}$, *DA2*){ (5.26)
  patch = ϕ;
  list_a, with triangle vertex pairs, is initialized with the pairs ($T_{i1,j}$, $V_{max\_nr1,j}$) and ($T_{i2,j}$, $V_{max\_nr2,j}$);
  vertex $V_{max\_nr1,j}$ is associated with axis *DA1*;
  vertex $V_{max\_nr2,j}$ is associated with axis *DA2*;
  do {
    list_b, with triangle vertex pairs, is emptied;
    for each pair ($T_{i3,j}$, $V_{k1,j}$) in *list_a* {
      sv = *set_vertices*($T_{i3,j}$, $V_{k1,j}$);
      st = *set_triangles*($T_{i3,j}$, $V_{k1,j}$)
      sv -= *patch*;
      all vertices in the state *SF* are removed from *sv*;
      take the growth axis *DAXIS* associated with vertex $V_{k1,j}$;
      for each vertex $V_{k2,j}$) in *sv*{
        associate vertex $V_{k2,j}$) with growth axis *DAXIS*;
        take a triangle $T_{i4,j}$ from *st* in which the vertex $V_{k2,j}$ occurs;
        add the pair ($T_{i4,j}$, $V_{k2,j}$) to *list_b*;
      }
      patch += *sv*;
      copy *list_b* to *list_a*;
    }
  } **while** (*list_b* is not empty);
} end *split_patch*

Dichotomous branching

In the algorithms (5.25) and (5.26) it is assumed that a maximum of two new growth axes can be found and *split_patch* will lead to the splitting of the original patch into two patches with distinct growth axes and lead to dichotomous branching. There is no reason to limit these algorithms to a maximum of two growth axes: it is only for simplicity reasons that the displayed algorithms are restricted to two growth axes. Both algorithms can be extended more or less straightforwardly to a larger number of growth axes. In the actual objects (see for example the sponge *Haliclona oculata* in Fig. 4.15) higher order branching can be observed quite often. In the actual objects the sites where material is added to the organism can be separated simultaneously into more than two patches where the addition of material continues.

5.6.6 The Coherence Conserving Rules

Until now two types of coherence conserving rules (the first and second *3D insertion rule*) were used, in which new longitudinal elements were inserted as soon as the tangential elements become too large. The type of rule being applied depends on whether triangles enlarge equally in all directions (isotropic dilation) or enlarge more in one direction than another (anisotropic dilation). In a branching object generated with the construction from Fig. 5.19F and displayed in Fig. 5.29, the reverse situation can also occur. In the branching process fertile vertices may become enclosed between the other vertices and situations can occur where there is not enough space for the formation of triangle edges larger than $0.5s$ (5.9). In order to resolve this situation it is necessary to delete longitudinal lines in the layered structure. This deletion can also be observed in Fig. 5.14, when $layer(2)$ is thought to be constructed upon $layer(3)$ and $layer(1)$ upon $layer(2)$. In Table 5.3 it can be seen that some of the longitudinal lines (for example the line between the vertices $V_{f,3}$ and $V_{f,2}$ in Fig. 5.14) disappear in the construction of $layer(1)$ upon $layer(2)$. The deletion of longitudinal elements should be applied while preserving the penta-hexagonal organization of the tessellation.

In Fig. 5.30 the deletion and insertion rules are displayed. The second *3D deletion rule* is applicable in the anisotropic case where in successive growth steps the size of tangential elements especially in one direction decreases. In the branching object depicted in Fig. 5.29 this situation occurs between the branches. In the actual objects this deletion can be observed in a longitudinal section, as shown in Fig. 3.11B, between the branches. Without the deletion of "non-fitting" vertices a skeleton would emerge in which the triangles become arbitrarily small, which is never found in reality (see Fig. 5.8).

3D deletion rule 2

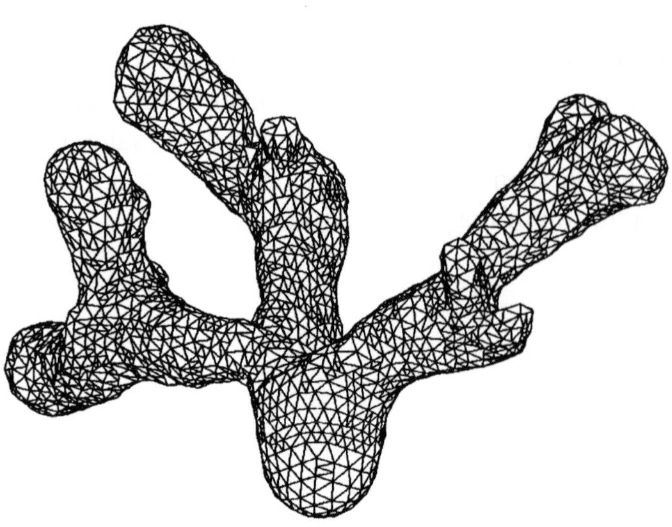

Fig. 5.29. A branching object generated with the construction from Fig. 5.19F

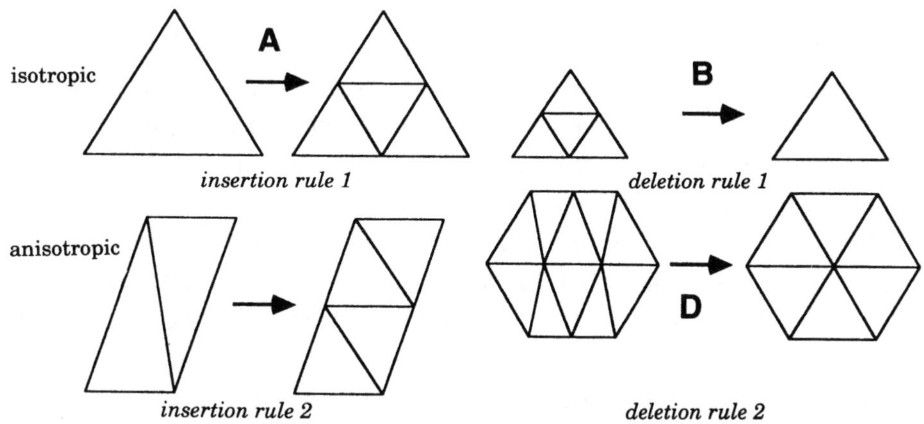

Fig. 5.30. Insertion and deletion rules for isotropic and anisotropic growth functions

3D deletion rule 1

The first *3D deletion rule* is only applicable in the isotropic case, where the edges decrease in size equally in all directions. The first *3D deletion rule* is exactly the reverse operation of the first *3D insertion rule*. An example of the case that a cluster of triangles emerges, that decrease in size equally in all directions in successive iteration steps, is the construction of a cup-formed object. At the bottom of the cup this cluster of too-small triangles evolves. This cup form is a common growth form among sponges (see van Soest 1989). A cup-formed object can be simulated by using an isotropic growth function (3.11) with a plateau with maximum values. This function will yield a club-shaped object, as shown in Fig. 5.19C. When this function is used instead of $f(\alpha, \beta)$ in (5.22) the club-shaped object will transform into a cup form at the moment max_curv in (3.14) is exceeded.

Coherence conserving rules in branching objects

In the branching objects which will be shown in the following sections, the coherence conserving rules A, C and D from Fig. 5.30 are used. In the experiments the tangential elements were allowed to vary within the range $0.5s..1.5s$. Edges below and above this range were removed using the insertion and deletion rules. In the objects it was nearly possible to create a tessellation satisfying this range, but in the simulations a residue of about 20 % of triangles with one or more edges below $0.5s$ was left over. It is still an open problem whether this condition can be satisfied completely in a tessellation where only 4-, 5- and 6-valent vertices are used.

5.6.7 More Evolved Branching Objects and Collision Detection

The Model. In the branching object shown in Fig. 5.29, there is no mechanism which prevents branches from growing through each other. To avoid physically impossible situations it is necessary to include an anti-collision rule. In the 3D models collision detection is done in a similar way as in the *avoidance rule* in the 2D models (see Sect. 3.6.7). In 3D collision detection, basically all vertices in the state F have to be compared with all vertices and this comparison takes O(number of vertices in the state F × number of vertices) computation time. Especially for large objects it is worthwhile to make some improvements in the efficiency of the collision detection algorithm and to try to reduce the number of comparisons.

Avoidance rule

Collision detection algorithm

In the collision detection algorithm in (5.27) in the first step the bounding box is determined that encloses the object. The bounding box is subdivided into $nsub\_boxes^3$ sub-boxes. After this a list is created with the sub-boxes and for each vertex $V_{k1,j}$ it is determined in which sub-box the vertex is located. Each sub-box in the list is filled with indices of vertices that are located in the sub-box. After this operation the number of comparisons can be reduced considerably. When the vertex is in state F, it only has to be compared with the vertices in sub_box containing $V_{k1,j}$. The immediate neighbours of $V_{k1,j}$, which are in the set $disc$, are excluded from the test. As soon as the distance d between $V_{k1,j}$ and another vertex $V_{k2,j}$ becomes too short and drops below the threshold min_dist, the state of the considered vertex is set to SF. This ensures that no new longitudinal elements are constructed upon this vertex and that self-intersections are prevented.

```
collision_detection(T, V){                                        (5.27)
  the upper, right, back and the lower, left, front corner of
  the box enclosing all vertices in V are determined;
  a list is initialized with nsub_boxes^3 sub-boxes;
  for each vertex V_{k1,j} ∈ V{
    the sub_box in which V_{k1,j} is located is determined;
    the index k1 is added to sub_box;
  }

  for each vertex V_{k1,j} ∈ V and its corresponding
  triangle T_{i1,j} ∈ T {
    if ( V_{k1,j} is in the state F){
      disc = sv = set_vertices( T_{i1,j}, V_{k1,j} );
      st = set_triangles( T_{i1,j}, V_{k1,j} );
      for each vertex V_{k2,j} ∈ sv {
        take a triangle T_{i2,j} from st in which V_{k2,j} occurs;
        disc += set_vertices( T_{i2,j}, V_{k2,j} );
      }
```

```
            the sub_box from the list with sub-boxes in which V_{k1,j}
            is located is determined;
            for each vertex (V_{k2,j} ∈ sub_box) && (V_{k2,j} ∉ disc) {
                d = || V_{k1,j}, V_{k2,j} ||;
                if ( d < min_dist ){
                    V_{k1,j} is set to the state SF;
                }
            }
        }
    }
} end collision_detection
```

The replacement system

The complete replacement system, including collision detection, is shown in a summarized form in (5.28). In this replacement system the exact description of the coherence conserving rules from Sect. 5.6.6 is left out for simplicity reasons.

$$
\begin{aligned}
initiator \quad = \quad & \mathcal{V} = \{(V_{1,1}, F, (0,1,1)); \cdots (V_{42,1}, F, (0,1,1)); \} \\
& \mathcal{T} = \{T_{1,1}; \cdots T_{80,1}; \} \\
generator \quad & (V_{a,j}, F, prev\_DA); \rightarrow \\
& \text{if } (l > inhibition\_level) \text{ then} \\
& (V_{a,j}, NF, prev\_DA); (V_{a,j+1}, F, prev\_DA); \\
& \text{else } (V_{a,j}, NF, prev\_DA); (V_{a,j+1}, SF, prev\_DA); \\
& (V_{b,j}, F, prev\_DA); \rightarrow \\
& \text{if } (l > inhibition\_level) \text{ then} \\
& (V_{b,j}, NF, prev\_DA); V_{b,j+1}, F, prev\_DA); \\
& \text{else } (V_{b,j}, NF, prev\_DA); (V_{b,j+1}, SF, prev\_DA); \\
& (V_{c,j}, F, prev\_DA); \rightarrow \\
& \text{if } (l > inhibition\_level) \text{ then} \\
& (V_{c,j}, NF, prev\_DA); (c_{b,j+1}, F, prev\_DA); \\
& \text{else } (V_{c,j}, NF, prev\_DA); (V_{c,j+1}, SF, prev\_DA); \\
& (V_{k1,j}, NF, prev\_DA); \rightarrow (V_{k1,j}, NF, prev\_DA); \\
& (V_{k2,j}, SF, prev\_DA); \rightarrow (V_{k2,j}, SF, prev\_DA); \\
& T_{i,j} = ((V_{a,j}, V_{b,j}, V_{b,j}), \cdots) \\
& T_{i,j+1} = ((V_{a,j+1}, V_{b,j+1}, V_{c,j+1}), \cdots) \\
& T_{i,j}; \rightarrow T_{i,j}; T_{i,j+1};
\end{aligned}
\tag{5.28}
$$

$$
\begin{aligned}
3D\ insertion & \\
rule\ 1 &= \cdots \\
3D\ insertion & \\
rule\ 2 &= \cdots \\
3D\ deletion & \\
rule\ 2 &= \cdots \\
3D\ association & \\
rule &= prev\_DA- > new\_DA; \\
3D\ anti-collision & \\
rule &= \text{if } (\|V_{k1,j}, V_{k2,j}\| < min\_dist) \text{ then} \\
& \quad (V_{k1,j}, F, new\_DA); \rightarrow (V_{k1,j}, SF, new\_DA);
\end{aligned}
$$

3D anti-collision rule

The effect of introducing the *3D anti-collision rule* is displayed in Fig. 5.31A and B. In this figure some evolved branching objects are shown that result after 50 iteration steps. The surfaces of both objects are visualized using Gouraud shading. (a description of this visualization technique can be found in Foley et al. 1990). In object A the anti-collision rule is not applied and the branches are growing through each other. When the rule is included in the iteration process it results in the thin-branching object B. In this object it can be seen that the branches are formed more or less in one plane (the plane with $\beta = 0$ in Fig. 5.24) as a result of using $f(\alpha, \beta)$ with the largest widening effect for $\beta = 0$.

Plate-like branches

In Fig. 5.32 the effect is shown of increasing the value of max_curv in the normalization function (3.14) applied in (5.22). The result is a more irregular object with a higher degree of branching and plate-like branches. The surfaces of the objects in Fig. 5.32 are visualized by Gouraud shading and ray-tracing[1].

In Fig. 5.31A and B it can be seen that the addition of one rule completely changes the overall branching pattern. It can be demonstrated that even for slight changes in a parameter value in the simulation model the overall branching pattern in the objects changes totally. In (5.29) the model parameter max_curv is determined by a function that uses the number of iteration steps it_step as an argument.

$$max\_curv = s \cdot \cos(start\_pt + \pi/28 \cdot it\_step) + 10s \qquad (5.29)$$

[1]The objects are visualized with the *rayshade* program written by Graig Kolb, developed at the University of Princeton.

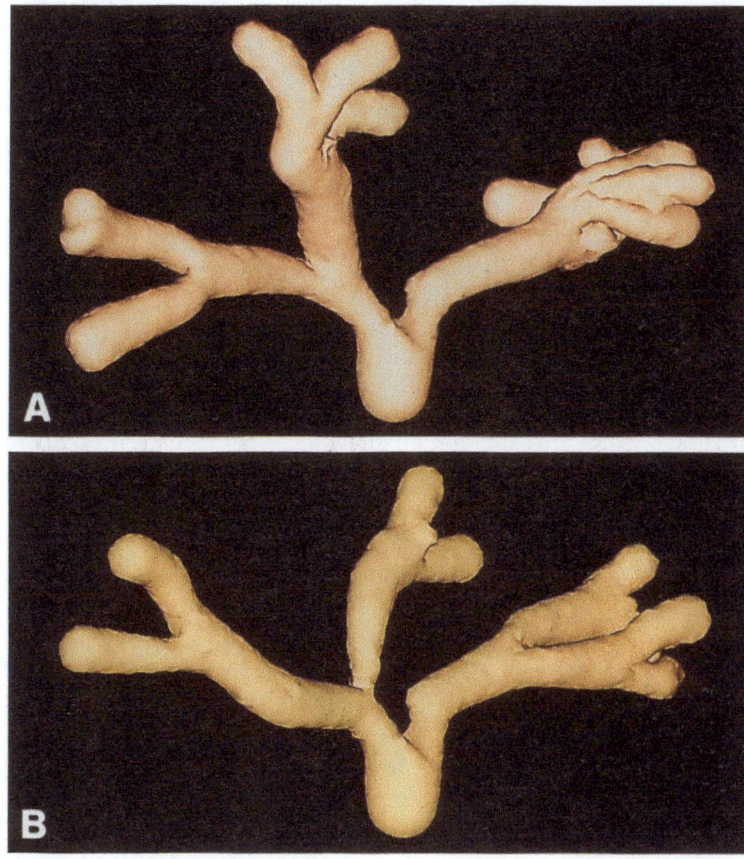

Fig. 5.31. Branching objects resulting from the replacement system in (5.28). The length l of new longitudinal elements was determined by (5.23) and (5.22). In the objects A and B the parameter *max_curv* was set to $10s$. In object A the *3D anti-collision rule* was not applied in the replacement system. The surfaces of the objects in A and B were visualized using Gouraud shading.

The result of this function is that *max_curv* oscillates slightly between the values $9s$ and $11s$ during the iteration process. A remarkable property of applying this oscillating value of *max_curv* in the iteration process is that it is very easy to generate a large set of different branching patterns by selecting different start points *start_pt* in (5.29).

A population of forms

In Fig. 5.33 four examples are displayed of objects generated with a slightly oscillating *max_curv* in the iteration process with different values of *start_pt* in (5.29). The four objects are visualized by applying Gouraud shading. Although the branching patterns are different in the four objects, the overall form of the objects can be described as thin-branching. Some of the morphological features (see Sect. 4.1) like the thickness of the branches remain the same. This experiment demonstrates that it is possible with this type of model of radiate accretive growth to create a population of forms which are different realizations of the same model.

The Biological Objects. The repulsive collision detection as applied in (5.27) corresponds with the *avoidance rule* as discussed in Sect. 3.6.7. In

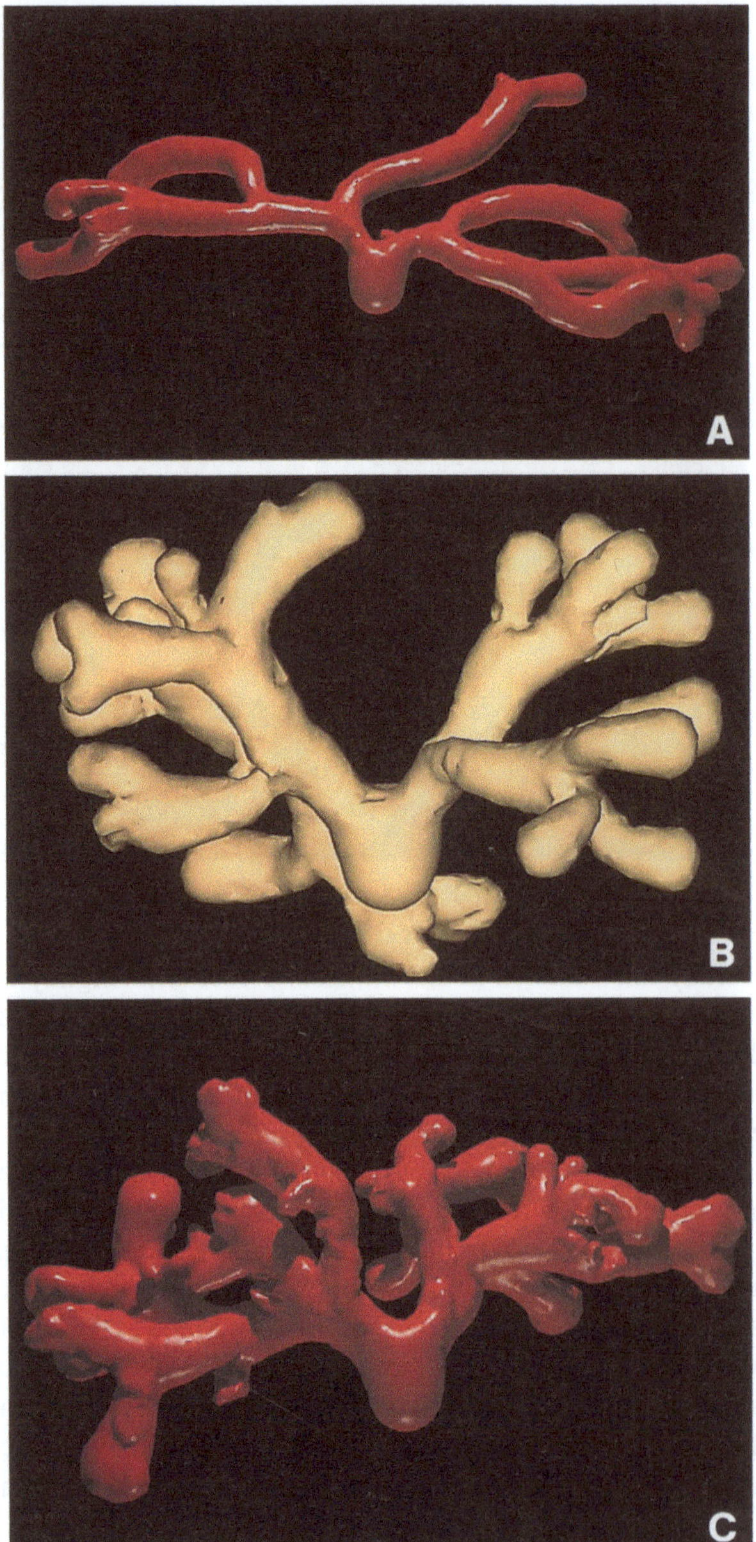

Fig. 5.32. Branching objects resulting from the replacement system in (5.28). The length l of new longitudinal elements was determined by (5.23) and (5.22). In A a thin-branching object is depicted, this object was generated in 90 iteration steps and the parameter max_curv was set to $10s$. In B the parameter max_curv was set to $12s$, in object C max_curv is set to the value $16s$. In B, and C the objects were generated in 50 iteration steps. The surface of the object in B was visualized using Gouraud shading, the surfaces of the objects A and C were visualized by ray-tracing.

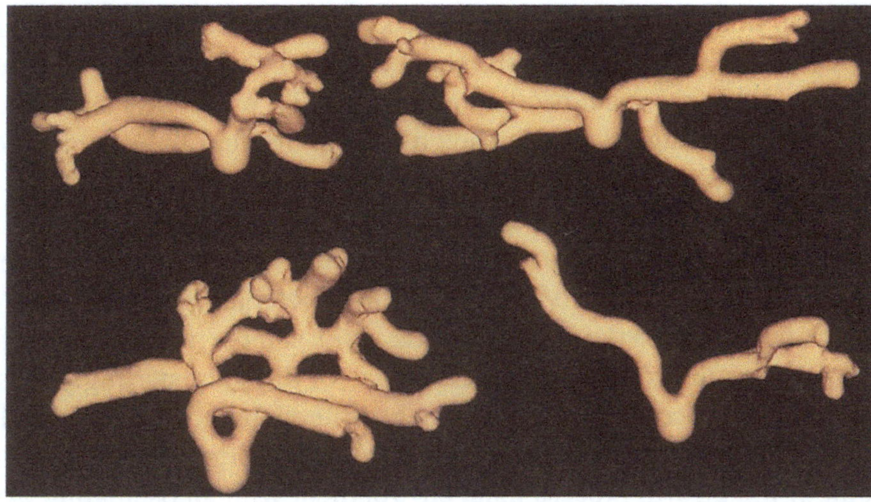

Fig. 5.33. Population of branching objects resulting from the replacement system in (5.28) after 80 iterations. The length l of new longitudinal elements was determined by (5.23) and (5.22). In all objects the parameter max_curv was determined by (5.29), $start\_pt$ was set respectively to the values 3, 5, 9 and 13. The surface of the objects was visualized by using Goureud shading.

some organisms the threshold min_dist is a specific distance; in the 3D models min_dist was set to the constant s.

In the following sections about the influence of the physical environment on the growth process, it will be demonstrated that it is possible to skip the *3D anti-collision rule* and to model the growth process where collisions are prevented in a more "natural" way.

A population of Haliclona oculata

In Sect. 4.1 it was demonstrated that it is possible to arrange some of the morphological features, for example the the diameter of the largest circle a which fits within the contour of a branching organism before a branch splits into new branches (Fig. 4.1), along an environmental gradient. In Fig. 4.4 the diameter of this circle da as measured in different populations of *Haliclona oculata* is related to the exposure to water movement. A population of *Haliclona oculata* at a certain growth site will be characterized by morphological features such as da, although none of the growth forms of this population will be exactly the same.

Fluctuations in exposure to water movement

In reality an environmental parameter such as exposure to water movement will never remain constant during time. Marine organisms typically live in an environment with tidal rhythms: the rising and falling of the sea twice each lunar day, which is again superimposed on the spring and neap cycle with a period of approximately 14 days. From the measurements in Table 4.14 it can be derived that the growth velocity of *Haliclona oculata* is in the range 1.0–1.5 cm in a period of about 10 weeks. When the amount of water movement is related with the morphological feature da it is expected that the increase in water movement during spring tide and the decrease during neap tide will cause a fluctuation in the width of the branches. It is expected that with precise measurements the spring and neap tide oscillation can be retraced in the growth form. In the photographs on Fig. 3.3 of *Haliclona oculata*, especially in the thin-branching growth forms, a certain oscillation in the width of the branches can be observed. Although this idea should still be verified with field experiments and precise measurements, this could be a possible explanation for oscillations in width of the branches.

Oscillations in growth forms

In the simulation experiment discussed above (see Fig. 5.33) such an oscillation was introduced in the model parameter max_curv. In Fig. 4.3 the relation between the parameter max_curv and the morphological feature da is visualized in a plot. The choice of different starting points in the iteration process leads to completely different branching patterns. It is expected that the same effect occurs in the actual organisms: the starting point of the growth process in an environment with an oscillating exposure to water movement and also slight disturbances in the growth process will result in an infinite amount of possible growth forms.

5.6.8 A Model of the Influence of Light Intensity on the Growth Process

The Model. The simple light model of Sect. 3.7.1 can also be applied for the morphological simulation of a non-branching, autotrophic organism with radiate accretive growth. A simulation of the colony shape of the stony coral *Montastrea annularis* (see Fig. 3.5) can be realized by using the simple light model $L(\theta)$ in the *generator processing function* from (3.19). The same function as applied in 2D model can be used in the 3D model because of the radiate symmetry of the colony. The angle θ is the angle between the mean normal vector in a vertex (see Sect. 5.6.2) and the vertical. Two results of this construction are shown in Fig. 5.19G1 and G2. In object G1 a column-shaped form is simulated and in G2 the hemispherical shape which emerges when reflection from the bottom is included. This reflection is simulated by applying a large value for max_angle in (3.19). In the column and hemispherical forms of Fig. 5.19G only the first *3D insertion rule* is necessary to preserve the penta- and hexagonal organization of the 3D tessellation, since the influence of the light intensity works equally in all directions.

Column-shaped and hemispherical forms

The simple light models in Fig. 5.19G1 and G2 only work in the case when no branches are formed. A model which includes the formation of branches and the influence of local light intensities is given below.

Local light intensities and the formation of branches

$$l = \begin{cases} s \cdot local\_light\_intensity \cdot curv\_index \text{ for} \\ local\_light\_intensity \cdot curv\_index > inhibition\_level \\ 0.0 \text{ for } local\_light\_intensity \cdot curv\_index \leq inhibition\_level \end{cases} \quad (5.30)$$

In this *generator processing function* a combination is made of (5.22) and the *local_light_intensity*; this light intensity in a vertex is no longer only determined by the angle θ as in (3.19). The *local_light_intensity* is a value determined by $L2(\theta)$ given in (5.31) and is also influenced by cast shadows.

Cast shadows

$$L2(\theta) = (1 - rest\_term) \cdot \cos(\theta) + rest\_term \qquad (5.31)$$

These shadows of the upper branches decrease the *local_light_intensity* in the lower parts of the object. In the algorithm given in (5.32) this effect of cast shadows on the value of $L2(\theta)$ is included. The result of this algorithm is an estimation of *local_light_intensity* in each vertex $V_{k,j}$ of the object, which is a value in the range 0.0 .. 1.0. In the function $L2(\theta)$ a rest term (*rest_term*) is used which describes the diffuse reflection from the environment. This *rest_term* represents, in most cases, a small percentage of the maximum light intensity (1.0); in the simulations values for *rest_term* in the range 0.0..0.6 are used. The angle θ is the angle between the normal vector of a triangle and the light direction (the vertical).

Diffuse reflection from the environment

```
det_local_light_intensity( T, V, lattice ){                                    (5.32)
  step A:
  for each triangle T_i1,j ∈ T {
    the local_light_intensity on T_i1,j is determined using L2(θ) from (5.31);
    local_light_intensity is added to lattice sites in the state "occupied"; }
  step B (cast shadows are determined):
  for each lattice site lattice[i][j][lattice_size - 1] {
    k = lattice_size - 1;
    object_found = FALSE;
    do {
      if (lattice[i][j][k] ≥ "occupied"){
        lattice[i][j][k] += "illuminated";
        object_found = TRUE; }
      k− −;
    while( ! object_found );
  }
  step C (non-illuminated sites are determined):
  for each lattice site lattice[i][j][k] {
    if ((lattice[i][j][k] > "occupied") &&
    (lattice[i][j][k] < "occupied + illuminated")) lattice[i][j][k] = "occupied";
  }
```

```
step D ("illuminated" mark is removed):
for each lattice site lattice[i][j][k] {
   if (lattice[i][j][k] ≥ "occupied + illuminated")
      lattice[i][j][k] -= "illuminated";
}
step E:
for each vertex V_{k,j} ∈ V and its corresponding triangle T_{i1,j} ∈ T{
   st = set_triangles( T_{i1,j}, V_{k,j} );
   local_light_intensity = 0.0;
   for each triangle T_{i2,j} ∈ st {
      local_light_intensity += mean value of light intensities in
      set_lattice_sites( T_{i2,j} ); }
   local_light_intensity /= number of triangles in st;
   the vertex V_{k,j} is associated with local_light_intensity;
}
} end det_local_light_intensity
```

Combination of the geometric and lattice model

In the algorithm in (5.32) a combination of the geometric model, represented by the lists $\mathcal{T}$ and $\mathcal{V}$, and the lattice representation $lattice$ is used. The lattice representation is obtained by mapping (see Sects. 5.5.1 and 5.5.2) the geometric model on a lattice. In Fig. 5.34 the three stages of this mapping on a lattice are shown. In the first stage triangles on the surface of the geometric model of Fig. 5.34A are mapped on a lattice, resulting in the discrete surface representation of Fig. 5.34B. In the second stage this discrete surface representation is changed into a discrete solid representation shown in Fig. 5.34C. In the determination of the local light intensities a virtual lattice of 400^3 lattice sites is used. For each triangle $T_{i,j}$ it is now possible to determine a set:

$$set\_lattice\_sites(T_{i,j}) \tag{5.33}$$

The local light intensity in a vertex

which contains the states of lattice sites by which $T_{i,j}$ is represented in the lattice model. The local light intensity at each vertex of the geometric model is approximated in the algorithm using the lattice model. Each lattice site can be brought into the state "unoccupied" or "occupied". When a site is in the state "occupied" it can thereupon be brought into the states "occupied + *local_light_intensity*" and "occupied + *local_light_intensity* + illuminated". The states are numbers which can be added and subtracted and they can have the following values:

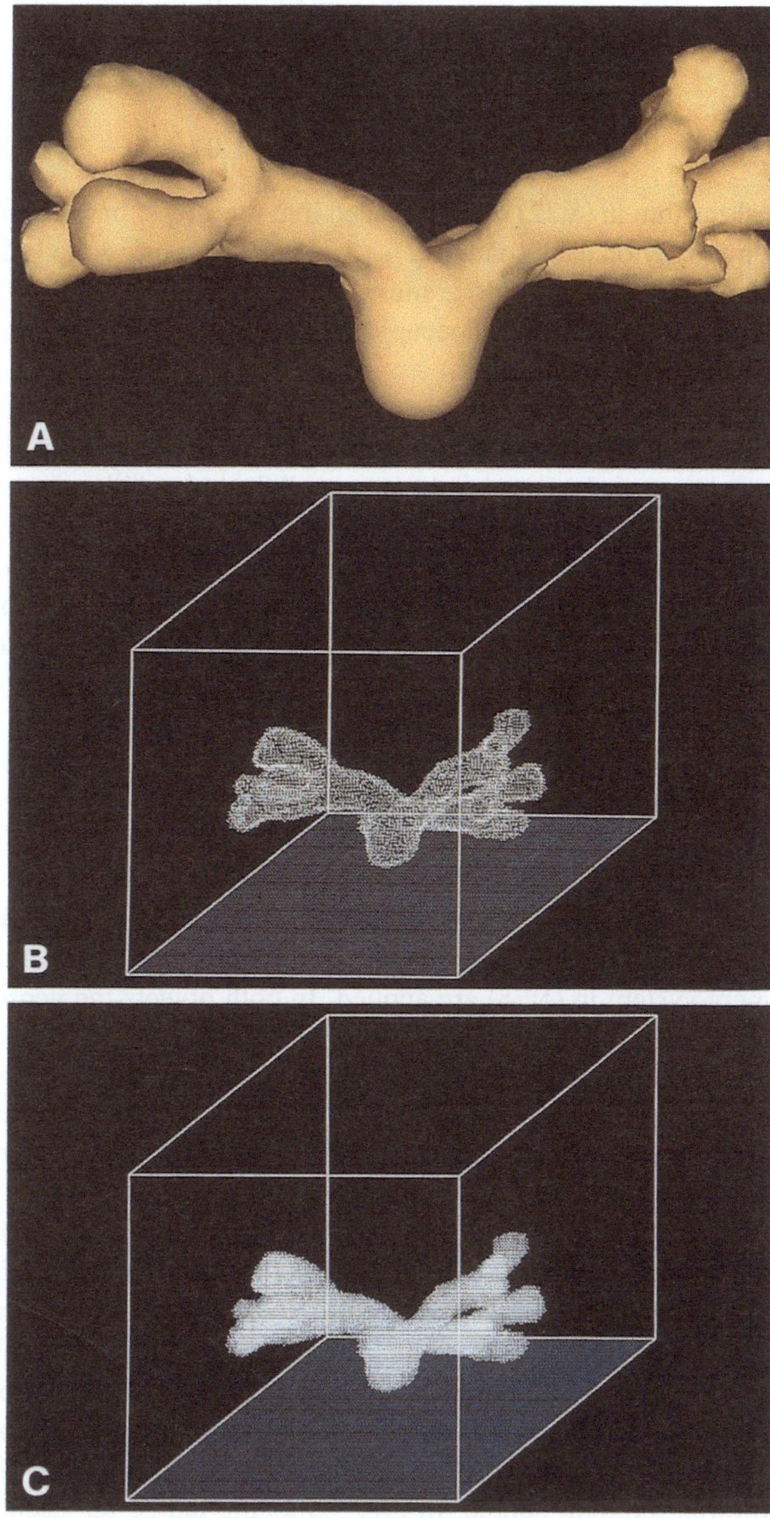

Fig. 5.34. Mapping of the geometric model on a lattice of 100^3 sites. In (A) a geometric model of a branching object is displayed. In (B) the discrete surface representation and in (C) the discrete solid representation of the object in (A) are shown.

$$\begin{aligned} \text{"}\mathit{unoccupied}\text{"} &= 0 \qquad (5.34)\\ \text{"}\mathit{occupied}\text{"} &> 1\\ \text{"}\mathit{illuminated}\text{"} &> 1\\ 0.0 \leq \mathit{local\_light\_intensity} &\leq 1.0 \end{aligned}$$

The light intensity function

The algorithm in (5.32) can be divided into five steps. In the first step the light intensity on each triangle $T_{i1,j}$ is determined using the light intensity function $L2(\theta)$. The lattice sites in the set $set\_lattice\_sites(T_{i1,j})$, which are already in the state "occupied" are brought into the state "occupied + *local_light_intensity*".

In the second step rays are casted from the top of the lattice to the bottom. When in a column of the lattice a site is encountered which is in the state "occupied + *local_light_intensity*" the value "illuminated" is added to the state of this site. The sites to which "illuminated" is added are visible from the top of the lattice and will be illuminated by a parallel light source with a light direction which corresponds to the vertical. The shaded sites will be after step B in the state "occupied + *local_light_intensity*".

Shaded sites

In step C the *local_light_intensity* of the shaded sites is set to the value zero. In step D the marker "illuminated" is removed. In the final step E for each vertex $V_{k,j}$ the mean value of *local_light_intensity* in the lattice sites situated in the triangles around $V_{k,j}$ is taken. The result is an estimation of *local_light_intensity* in the vertex $V_{k,j}$; this value can be associated with the vertex as an attribute in the replacement system.

In Fig. 5.35A and B two more evolved examples are displayed of the models where the *generator-processing function* in (5.30) and the replacement system in (5.28) are applied. In both objects the geometric model is visualized with Gouraud shading and in Fig. 5.35C the lattice version of object B is shown. In object C the lattice sites which are situated on the surface of the object are shown, and the local light intensity is visualized by the colours gray (zero light intensity) and red, where the brightness indicates the light intensity.

Visualization of the local light intensities

In the model displayed in Fig. 5.35 the growth process is influenced by the combination $local\_light\_intensity.curv\_index$ in (5.32). In Fig. 5.36 some more evolved examples are shown of models where the length of new longitudinal elements is determined by the combination $f(\alpha, \beta).local\_light\_intensity.curv\_index$ ($f(\alpha, \beta)$ is described in (5.20)). In A the developmental sequence is shown as the object depicted in B emerges. In this picture it can be seen that the branches are developing towards the light source. In Fig. 5.36C the lattice representation of the object in B is shown; from this picture the local light distribution over the object can be derived. In this model the contribution of

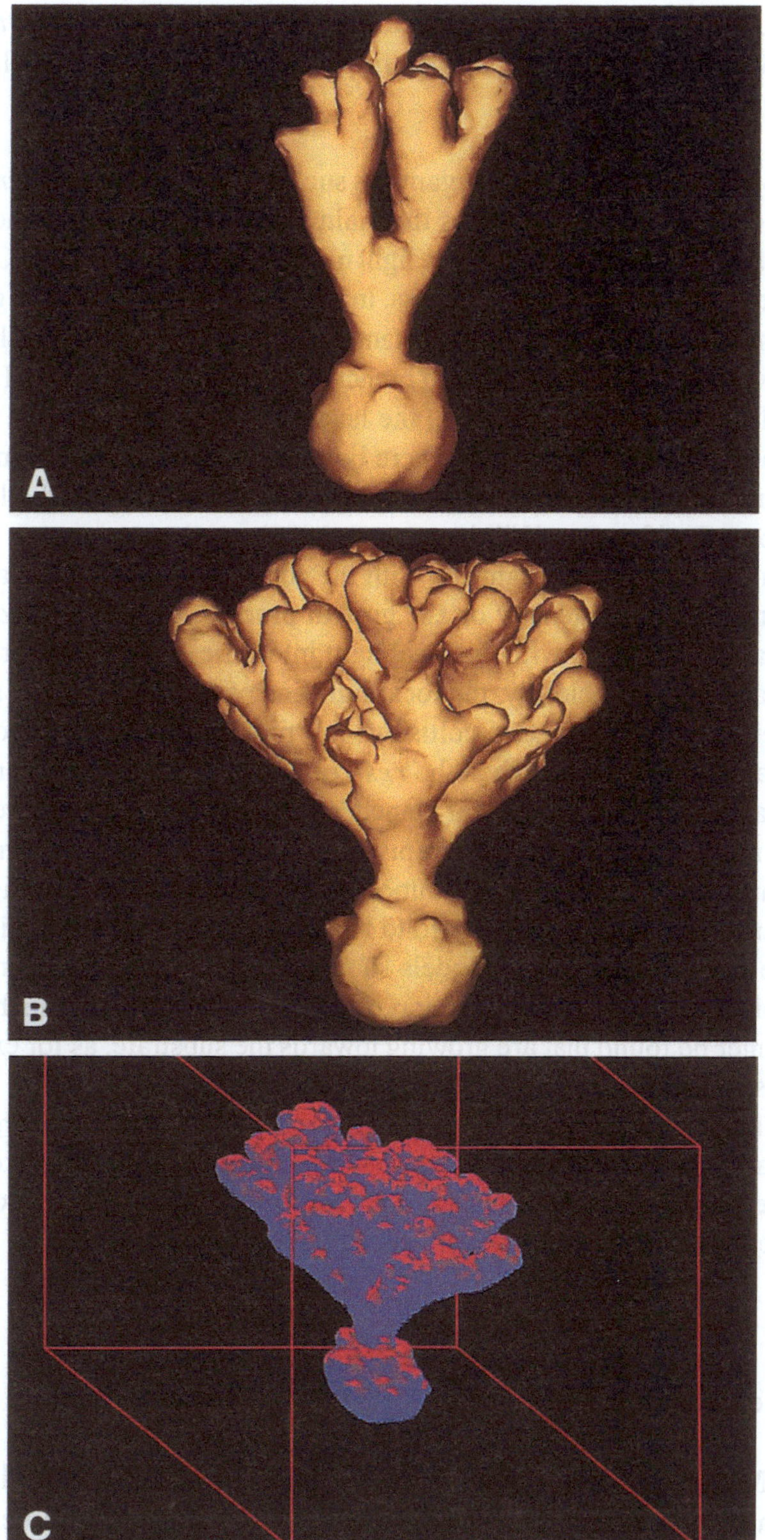

Fig. 5.35. Branching objects resulting from the replacement system in (5.28) after 80 iterations. The length l of new longitudinal elements was determined by (5.30). In all objects the parameter max_curv was set to the value $10s$. The parameter $rest\_term$ in the light intensity function $L2(\theta)$ used in (5.32) was set respectively to the values 0.0 and 0.3 in the objects A and B. Both objects are visualized with Gouraud shading. In object C the lattice version of object B is displayed and the local light intensity is visualized by the colours gray (zero light intensity) and red, where the brightness indicates the light intensity.

local_light_intensity can be controlled with the parameter *rest_term* in (5.31); for the depicted objects this parameter was set to the relatively high value 0.6.

The Biological Objects. In organisms such as *Montastrea annularis* (see also Sect. 3.7.1), where light is the main energy source and without the formation of branches, the simple light models in Fig. 5.19G1 and G2 may serve to simulate the growth process. In many organisms where the growth process is influenced for an important part by the distribution of light in the environment, the formation of branches is found (see Fig. 3.8). In these organisms often light as well as the heterotrophic energy source (filter-feeding) are used. Organisms where both energy sources are significant are for example the Scleractinian *Acropora palmata* (see Bythell 1988) and some Porifera (see Wilkinson et al. 1988).

Organisms without branch formation

In the models of organisms with the formation of branches and the influence of light intensity on the growth process, the *3D anti-collision rule* can be modelled with the local light intensities. In the case branches approach each too closely, the local light intensity will decrease as well as the growth velocity and collisions are prevented. This anti-collision mechanism only works when the *rest_term* in (5.31) is set to zero. Another aspect of the formation of branches which is captured quite well with these models is positive phototropism. In the pictures of organisms with a significant autotrophic contribution to the energy intake (see Fig. 3.8), a clear tendency can be observed to form branches towards the direction of the light source. In the models shown in Fig. 5.35A and B branches are formed mainly in the direction of the light source and branches can no longer be found that are growing towards the substrate, as for example occurs in the models shown in Figs. 5.32 and 5.33.

Organisms with the formation of branches

Positive phototropism

The objects shown in Fig. 5.36 can be used as a model of an organism with an anisotropic growth process, where the growth velocities depend on the position of the organism with respect to the flow direction, and where the growth process is influenced by local light intensities.

5.6.9 A Model of the Influence of Nutrient Distribution on the Growth Process

The Model. In Sect. 3.7.2 it was discussed that the nutrient distribution around an organism, under sheltered conditions, can be described with the Laplace equation (2.7). The same algorithm as applied in the 2D model (3.21) can, in an extended version, be used to determine the

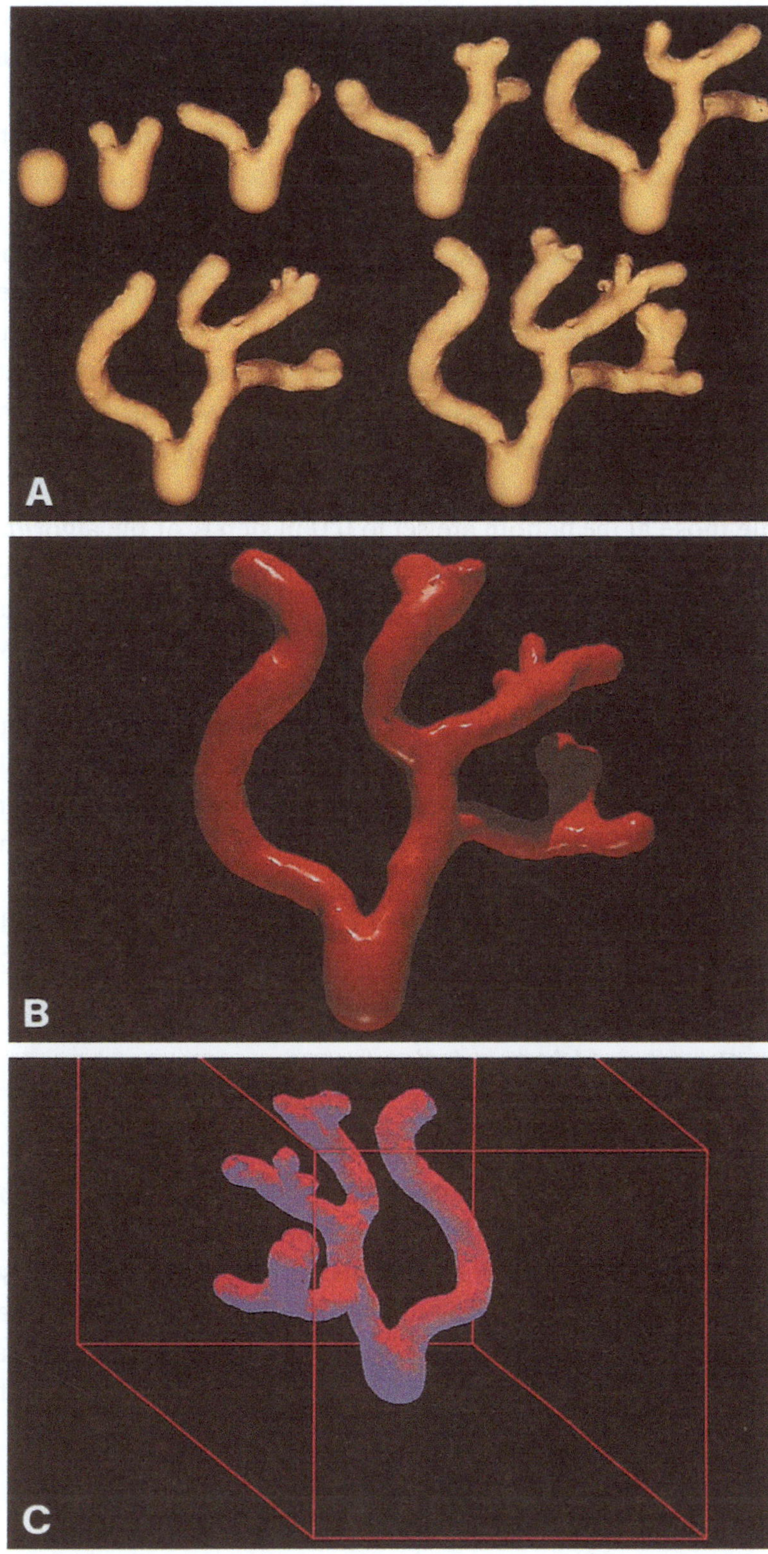

Fig. 5.36. Objects resulting from the replacement system in (5.28) after 80 iterations. The length l of new longitudinal elements was determined by (5.30), in this function the combination $local\_light\_intensity \cdot curv\_index$ was replaced by $f(\alpha, \beta) \cdot local\_light\_intensity \cdot curv\_index$. The parameter max_curv was set to the value $10s$. The parameter $rest\_term$ in the light intensity function $L2(\theta)$ used in (5.32) was set to the value 0.6. In picture A the developmental sequence of the object displayed in B is shown. The objects in A are visualized with Gouraud shading, object B by ray-tracing. In object C the lattice version of object B is displayed, the local light intensity is visualized by the colours gray (zero light intensity) and red, where the brightness indicates the light intensity.

nutrient distribution around the 3D object. The extension of the algorithm is straightforward. The mapping of the object into the 3D lattice can be done with the methods described in Sect. 5.5.1

Laplace equation

A few serious practical problems occur in step C in the algorithm (3.21) where the Laplace equation is solved. In Sect. 5.5.1 it was mentioned that it is often impossible to allocate a lattice with 1000^3 sites. This limitation can be overcome by using virtual lattices, as described in Sect. 5.5.2. The next problem is that when two of these virtual lattices are used, the convergence in the 3D extension of (3.21) step C is still unacceptably slow[2]. This problem was avoided by using two normal lattices of 100^3 sites whether this decreased resolution gives an inaccurate approximation of the nutrient distribution around the object is still an open problem. In the 3D models the lattice sites at the top are set to the value 1.0, while those at the bottom are set to 0.0 in step C of the algorithm.

Local nutrient gradient

In the 3D version of the nutrient model an estimation is done of the local nutrient gradient at each vertex $V_{k,j}$ in the state F. For this purpose an edge *probe* is drawn in step D of the 3D version of (3.21), with a 3D version of the Bresenham algorithm (see Wijkstra 1991). The edge starts at the vertex $V_{k,j}$ and has the same direction as the longitudinal element $(V_{k,j}, V_{k,j-1})$ and points into the environment around the object. The gradient is estimated using the values of lattice sites in *probe*; in the estimation the relation between the local field and the concentration $k(c)$ (3.23) was assumed.

The influence of the local nutrient concentration can be included in the model by applying a combination of $k(c)$ and $curv\_index$ (5.22) in the *generator processing function*:

$$l = \begin{cases} s \cdot k(c).curv\_index \text{ for} \\ k(c) \cdot curv\_index > inhibition\_level \\ 0.0 \text{ for } k(c) \cdot curv\_index \leq inhibition\_level \end{cases} \tag{5.35}$$

Sections through the lattice model

Two views of a more evolved example in which this function was applied are shown in Fig. 5.37A and B. In Fig. 5.38 sections through the 100^3 lattice model are shown. In these sections the object itself is displayed in red, and the basins of equal nutrient concentration around are visualized in alternating black and coloured regions. The colour gradually changes from white to blue; this colour shift indicates an increasing nutrient concentration.

[2]To obtain a branching object, on a sparc 4 workstation, in 80 iteration steps with this method would take approximately 186 days!

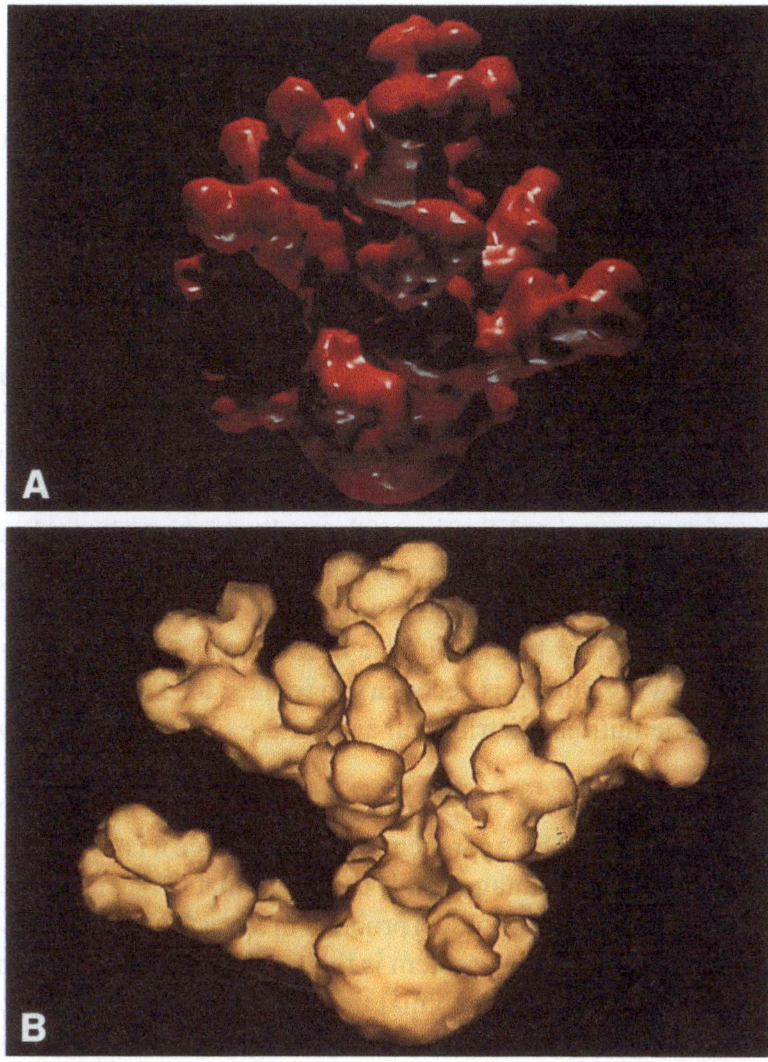

Fig. 5.37. Two views of an object resulting from the replacement system in (5.28) after 80 iterations. Object A is visualized by ray-tracing, object B by applying Gouraud shading. The length l of new longitudinal elements was determined by (5.35). The parameter max_curv was set to the value $10s$, while η in (3.23) was set to 0.5.

The Biological Objects. The model of a branching object in Fig. 5.37, where the nutrient distribution is mimicked by a diffusion process, exhibits several properties which correspond to the growth process of a heterotrophic organism under sheltered conditions (see also the paragraph on the biological objects in Sect. 3.7.2). In this model the branches are formed towards the nutrient source, which can be used to simulate negative substrate-tropism. Collisions in the model are prevented since the branches compete for the same nutrients, whose concentration locally between the branches becomes near zero so that growth is suppressed.

Negative substrate tropism

5.7 Conclusions and Restrictions of the Presented 3D Models

The 3D models of radiate accretive growth presented in this chapter are a starting point for a larger system of 3D models. With the presented models the problems discussed in Sect. 3.8, on the restrictions of the 2D model of radiate accretive growth, are partly solved.

Nutrient distribution model

The Laplacian model of the nutrient distribution around the growing object, presented in Sect. 5.6.9, is only applicable to mimicking the nutrient distribution under sheltered conditions. A more refined model requires the inclusion of a model of the flow and the corresponding nutrient supply around the object. In the flattened model (Fig. 3.17E), the growth velocities are unequal on the object and depend on the position of the object with respect to the flow direction (Figs. 5.23 and 5.26). This assumption is a strong simplification of the reality and is only applicable for simple objects. In an actual object the flow patterns around the organism can become highly complex and are influenced strongly by the form of the organism. In order to simulate the nutrient supply in an object which grows in a moving fluid, it is necessary to replace the Laplace equation by a model of the nutrient distribution around the object which allows drift of nutrients.

Estimation of the local curvature

In Sect. 5.6.5 a rather simple approach was used to estimate the local curvature in a vertex. This approach was chosen since it is computationally a fast method. A disadvantage of this method is that it may introduce noise in the model. This is a well-known problem in the estimation of curvatures. A possible solution of this problem is to use polynomial approximations, in which the best fitting polynomial through a set of points using a least squares method is approximated (see also Terzopolous 1986, Besl and Jain 1986, Lim et al. 1990)

In a model of a branching organism with a combination of an autotrophic and heterotrophic metabolism, a light and a nutrient distribution model are necessary. In a simulation of an organism with a combined energy source the influence of both environmental parameters can be weighted. For some organisms, for example *Acropora palmata* (see Bythell 1988), the percentage of the contribution of the light source is known. Basically it is possible to construct models with such a combined energy source; a simple example of this was shown in Fig. 5.36.

Anastomosis and abrasion

Two more aspects of the growth which cannot be modelled with the presented 3D model are anastomosis and abrasion. Anastomosis (the process in which branches fuse) is a very characteristic (see also Sect. 3.8) phenomenon often found among marine sessile organisms. The inclusion

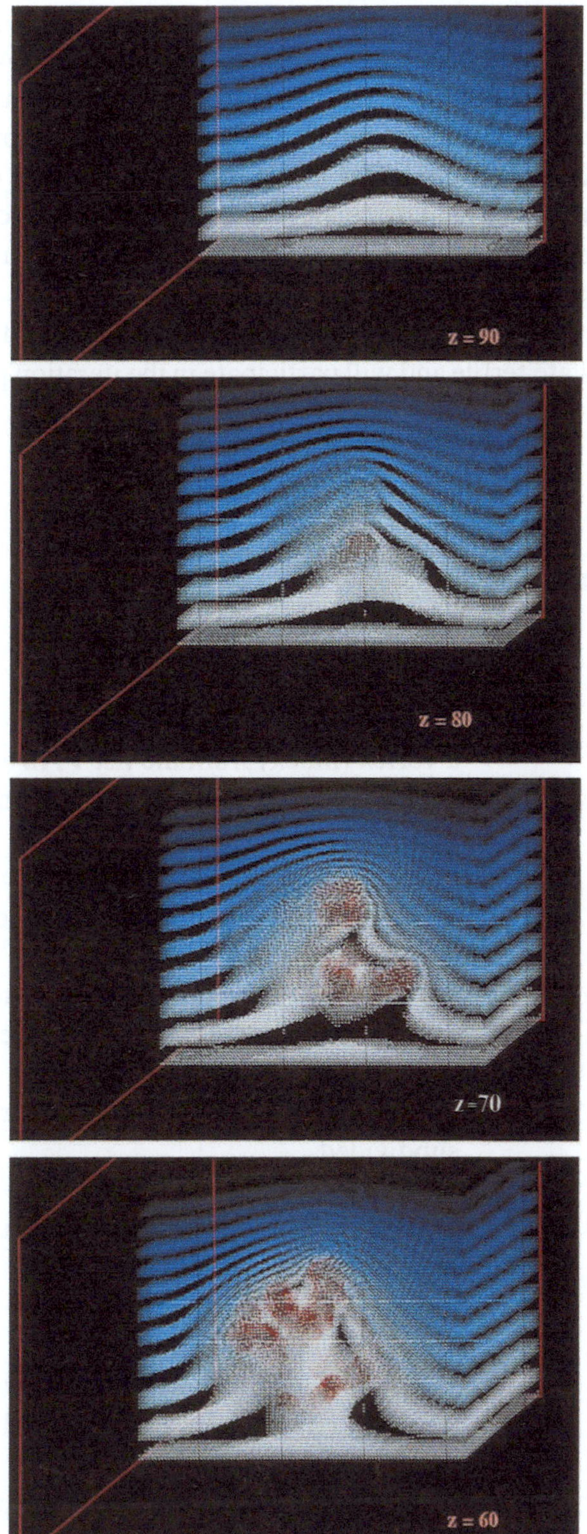

Fig. 5.38. Sections at different k-planes (indicated as the z-value in the pictures) through the lattice representation with 100^3 lattice sites, of the object. The object itself is displayed in red, and basins of equal concentration ranges are alternately visualized as black and coloured regions. The shift from blue to white indicates a depletion in nutrient.

Negative growth

of this effect will increase both the "realism" and the complexity of the generated forms significantly. The aspect of "negative growth" by abrasion will have the same consequence. The multi-layered structures shown in this chapter are suitable to model the process of abrasion. In a model of abrasion the outermost layers at the protrusions of the object can be assumed to have the highest chance of being removed from the object during the iteration process. The growth process can continue from these damaged sites on the object by secondary growth (see Sect. 3.6.3). In the 3D model fertile vertices in the state "non-active", at the damaged sites, are set to the state fertile and will participate in the iteration process again.

5.8 List of Symbols Used in Sects. 5.3 to 5.7

$\mathcal{V}$	vertex index list
V_i	a vertex (index) from the list $\mathcal{V}$, in a single-layered triangular tessellation
$\mathcal{C}$	coordinate list
$\mathcal{T}$	triangle list
T_i	a triangle from the list $\mathcal{T}$, in a single-layered triangular tessellation
$set\_triangles(T_i, V_i)$	set of triangles surrounding the vertex V_i
$set\_vertices(T_i, V_i)$	set of vertices surrounding the vertex V_i
s	basic size of a tangential element
$level$	the number of times a triangle is further subdivided
$null$	no reference to another triangle
n_i	vertices newly inserted in $\mathcal{V}$ when a triangle is subdivided
Ti	triangles newly inserted in $\mathcal{T}$ when a triangle is subdivided
$V_{i,j}$	vertex V_i from layer j in a layered tessellation
$T_{i,j}$	triangle T_i from layer j in a layered tessellation
F, NF, SF	a vertex can be respectively in the state: "fertile", "not-fertile", or "non-active"

l	length of a longitudinal element $(V_{k,j}, V_{k,j-1})$
$inhibition\_level$	threshold below which l becomes zero
α	angle between an axis of growth and the direction of a newly constructed longitudinal element
β	angle between the projection of the mean normal vector on the xz plane, in a vertex $V_{i,j}$ and the x-axis. The direction of the flow corresponds to the z-axis
$f(\alpha, \beta)$	function describing the deposition of a new layer of tangential and longitudinal elements
w	widening factor in $f(\alpha, \beta)$
max_curv	maximum value radius of curvature, expressed in units s
$h(rad\_curv)$	function which returns a normalized version of the radius of curvature
$layer(j)$	a list with triangles $T_{1,j}, T_{2,j}, ..$
$lattice$	the lattice representation of the geometric model
$lattice[i][j][k]$	a lattice site with coordinates i, j, k
"occupied"	lattice sites by which an object is represented
"unoccupied"	lattice sites that are not occupied by an object
"recently_occupied"	special marker used in the flood-fill algorithm (5.17)
"subdivided"	special marker used in the virtual lattices to indicate that a lattice site further subdivided
"illuminated"	special marker used in the local light intensity algorithm (5.32)
$lattice\_size$	maximum of the i, j, k-coordinates in $lattice$
$curv\_index$	index summarizing the normalized radii of curvature in a vertex $V_{k,j}$
low_norm_curv	the lowest value of the normalized radii of curvature in a vertex $V_{k,j}$
av_norm_curv	the average value of the normalized radii of curvature in a vertex $V_{k,j}$
$fertile\_patch$	patch of neighbouring vertices in the state F
$DAXIS$	vector with the direction of a growth axis
$prev\_DA$	vector with the direction of the growth axis before applying the *3D association rule*

new_DA	vector with the direction of the growth axis after applying the *3D association rule*
$nsub\_boxes$	the number of times an edge of the bounding box containing the object is subdivided in the collision detection algorithm in (5.27)
sub_box	one of the $nsub\_boxes^3$ sub-boxes in which the bounding box containing the object is subdivided
min_dist	minimum distance between a fertile vertex $V_{k1,j}$ and the other vertices $V_{k2,j}$ in state F or SF used in the collision detection algorithm in (5.27), expressed in units s
it_step	number of the iteration step
$start\_pt$	starting point of the function in (5.29) returning an oscillating max_curv
a	the largest circle which fits within the contour of a branching object or organism, before the branch splits into new branches
da	the diameter of the circle a
$L(\theta)$	function which returns the light intensity in a vertex $V_{k,j}$ in a non-branching object
θ	The angle between the mean normal vector in a vertex $V_{k,j}$ and the vertical
max_angle	maximum angle which a longitudinal element can make with the vertical
$local\_light\_intensity$	local light intensity in a vertex $V_{k,j}$ of a branching object
$L2(\theta)$	function which returns the light intensity in a triangle $T_{i,j}$, used in the local light intensity algorithm for a branching object in (5.32)
$rest\_term$	a term in $L2(\theta)$ that describes the contribution of the diffuse reflection from the environment
$set\_lattice\_sites$	set that contains the states of the lattice sites by which the triangle $T_{i,j}$ is represented
$probe$	an edge in the $lattice$ starting in the vertex $V_{k,j}$ with the direction of the longitudinal element ($V_{k,j}$,$V_{k,j-1}$) and pointing in the environment of the object

$k(c)$	function which represents the influence of the nutrient distribution on the growth process
η	exponent describing the relation between the local field and the concentration c

6 Final Conclusions

6.1 The 2D and 3D Simulation Models

In Chaps. 3, 4 and 5 it was demonstrated that the radiate accretive growth can be simulated in 2D and 3D with a geometric model. In these models growth is described as an iterative process in which the growing object is represented by a geometrical object. In the iteration process a set of rules is applied which can be divided into two types: rules representing the internal properties of the growing object and rules which represent the influence of the environment on the growth process.

For a branching organism with radiate accretive growth the internal rules can be summarized as:

a) *generator*, the geometric construction describing how new skeleton elements are added to a preceding growth stage.
b) A *generator-processing function* describing the secretion of elements over the surface of the object ($f(\alpha)$).
c) A generator-processing function which describes the limitations of the transport system of nutrients through the tissue of the organism ($h(rad\_curv$ in the 2D and $curv\_index$ in the 3D models).
d) *post-processing rules* describing that all elements are connected in a coherent structure (*insertion, continuity* and *deletion rule*).
e) A *post-processing rule* which represents the formation of new growth axes (*association rule*).

The external rules can be summarized as:

a) *post-processing rules* which represent geometric restrictions (anti-collision rules).
b) A *generator-processing function* representing disturbances of the growth process ($g(lowest\_value)$).
c) A *generator-processing function* which represents the influence of the nutrient distribution on the growth process ($k(c)$).

d) A *generator-processing function* which represents the influence of the light intensity on the growth process ($L(\theta)$ in the 2D and $local\_light\_intensity$ in the 3D models).

For many of the organisms only a subset of these rules are necessary; which rules are relevant is determined by species-specific properties (e.g. autotrophic or heterotrophic organisms, internal or external secretion of elements).

Radiate symmetry and 2D models

Some aspects of the growth process, such as the formation of thin-branching and plate-like forms, can be described with a 2D model. This is possible because in a radiate accretive growth process basically a structure is formed with a radiate symmetry. In Sect. 4.1.1 it is demonstrated that some aspects of a range of ecotypes found along a gradient of exposure to water movement can be simulated in a series of experiments where the thin-branching forms gradually transform into plate-like ones. It is also shown that there is a relation between the model parameters max_curv and $lowest\_value$ and the observed forms. Using this relation it is, in theory, possible to simulate a given growth form in a gradient of exposure to water movement.

Flattened forms and 3D models

Some aspects of the growth process, such as the formation of a flattened growth form and the possibility of branches to avoid each other in space, can only be simulated with a 3D model. In Sect. 5.6.5 a 3D model of a flattened growth with the formation of branches is shown. In this model a flattened form is generated by assuming a *generator-processing function* $f(\alpha, \beta)$, in which the secretion of elements also depends on the angle β with the flow direction. This is a drastic simplification of reality. In an improved version where also drift of nutrient is included, it may be possible to simulate the emergence of flattened forms in a more natural way. The 3D models exhibit a relatively much higher complexity and require several computationally expensive steps. The presented 3D models of radiate accretive growth are only a starting point for a more extensive system of 3D models.

Verification of the models

The result of this book is a 2D and 3D modelling system capable of modelling a growth process from which a large class of objects can emerge. The assumptions in the model were verified with experiments on the actual objects. These experiments show that the model has a predictive value. Some of the predictions can be done with the 2D model by using the radiate symmetry of the simulated organisms. A starting point for a more refined 3D model of radiate accretive growth is presented in Chap. 5. The general method for iterative geometric constructions, as presented in Chap. 2, can be used as a starting point for other classes of models, capable of simulating other types of growth processes. The methods presented in Chap. 4 can be used more in general for the comparison of growth forms as found among marine sessile organisms, and for the comparison of simulated forms and actual forms. The same type of experiments described in the same chapter may be used in further research on the emergence of growth and form in marine sessile organisms.

6.2 Application of the Simulation Models in Ecology

Insight into morphogenesis

The first aim in constructing morphological models of growth processes is to gain insight into the morphogenesis of organisms. Even relatively simple growth processes may lead to a surprisingly large variety in growth forms. A substantial part of the growth forms found in a coral reef may be covered, when some simplifications are assumed, by the 2D and 3D models presented in this book. In a more extensive approach these models can be extended by some more basic types of growth processes other than the radiate accretive growth process which is used as case study in this book. In such an approach it is necessary to describe the various aspects of these types of growth processes in formal rules. This leads to the identification of parameters responsible for certain features in the growth forms. Such an approach can be used for causal explanations in taxonomy. It becomes possible to predict which forms, for a given species with a species-specific architecture, can develop in the growth process (see Sect. 3.5). It can for example be predicted that a species with a skeleton where the elements are oriented randomly will usually develop quite irregular (often encrusting) growth forms.

Species-specific parameters

When comparing two related species, e.g. the sponges *Haliclona oculata* and *Haliclona simulans*, it is possible to identify the parameters which cause the species-specific differences in growth forms. As explained in Sect. 3.5, the aquiferous system in *Haliclona simulans* is more evolved, causing the emergence of a more globular and wide-column habit than in *Haliclona oculata*. A second architectural aspect which causes a species-specific difference in growth form is the organization of the tangential layers in both species. A consequence of the triangulate architecture in *Haliclona simulans* is that the overall growth form will be more rigid, while a penta-hexagonal organization, as found in *Haliclona oculata*, allows for a more flexible architecture. This more flexible structure resists strong water movements better and allows a growth form which can survive in a larger range of habitats with different rates of water movement. The more rigid architecture of *Haliclona simulans* leads to a more vulnerable structure, which will often develop a more creeping growth form. The original tree-like form is damaged and pushed on the substrate, and the tree-like form emerges again (compare the simulated form in Fig. 4.8). This phenomenon is described by Dauget (1991) as a reiterated model. The flexible skeleton architecture, as found in *Haliclona oculata*, is more apt to develop the typical tree-shape of this species.

Influence of the environment

Next to the species-specific parameters, the shape of an organism in a certain growth stage is determined by the influence of the environment. Many modular organisms show a clear response in growth form to the governing environmental conditions (see Sect. 3.3). In order to identify which aspects of the growth form are influenced by the environment, a combination of a morphological model with species-specific parameters and a model of the physical environment is very useful. In such a model it becomes possible to predict the possible range of ecotypes of a certain species.

Environmental gradients

The variety in forms, caused by the influence of the environment, can often be arranged along a gradient of a changing environmental parameter. In the case of more than one dominant environmental parameter a phase diagram can be composed (compare the phase diagram for DLA clusters in Ohgiwari et al. 1991). Each point in such a diagram corresponds to a certain setting of the environmental parameters and represents a certain realization of the growth form. In these response curves the self-similar aspect (e.g. width of branches, degree of branching, etc., see Fig. 4.4) or the fractal dimension (see Table 4.4) can be plotted against the value of the dominant environmental parameter.

Bio-monitoring

The response curves are useful for bio-monitoring purposes. From these curves it becomes possible to estimate the value of an environmental parameter from a given growth form. In Fig. 4.4 an estimation is made of the exposure to water movement using the growth form. In the paper of Jebram (1980) it is demonstrated for the bryozoan *Electra pilosa* that there is a relation between the colony shape and the nutrient supply. In the study of Bosence (1976) on coralline algae, it is demonstrated that these organisms show a clear response in the growth form to the exposure to water movement. The growth process of these algae can be modelled as a radiate accretive one. Species of these unattached coralline algae are an important part of the marine communities in tropical as well as temperate environments. These algae also form a significant part of the marine sediments. These features make growth forms of this algae group very suitable as biomonitors of recent and paleoenvironmental conditions.

The growth form is not only a reflection of the governing environmental parameters, but can also be used as a continuous registration of the values of these parameters. An impressive example of this can be found in Marden (1978), where a section through a coral is shown (comparable with the section shown in Fig. 3.9) in which the growth velocities are recorded in the period between 1620 and 1975. The fluctuations in growth velocities can be correlated with seasonal variations by combining such a registration with a simulation model of the growth process and the influence of the physical environment.

Changes in the environment

Sudden changes in the environment, causing a disturbance of the growth process, can also be detected in growth forms. In the transplantation experiments (see Sect. 4.2.3) it was demonstrated that on one of the three experimental sites growth was stunted by increased sedimentation. This type of disturbance might also lead to deviating growth forms, which do not correspond with the forms "normally" observed or with the simulated forms. In Fig. 4.7 the effect on the growth form of *Haliclona oculata* of a sudden change in the exposure to water movement is simulated. These predictions can explain deviating forms which can be demonstrated experimentally (Fig. 4.13) and can be found occasionally in the field (Fig. 4.15).

Toxic agents

Next to a sudden change in sedimentation or exposure to water movement, toxic agents may disturb the growth process. An example of a growth form in which the growth process was experimentally disturbed by a (toxic) agent[1] is displayed in Fig. 6.1. In this experiment a cup-shaped sponge was found. This form can be explained by assuming a partial death or suppression of the secreting cells which are relatively most exposed to the environment and are situated in the tip of the sponge. The secreting cells at the borders of the tip are less exposed to the environment and can continue to grow. The result is that along the borders of the local minimum (the cup) a series of local maxima arises. This form can be simulated with the clavate model (Fig. 5.19C). On the tip of club-shaped form the growth velocity will locally decrease when the radius of curvature exceeds a certain maximum or when growth is simply stopped by setting the fertile vertices to non-fertile. Around the local minimum branches will be formed. This type of form is also sometimes found in the field (Fig. 6.2) and a possible explanation is a temporary disturbance leading to a partial mortality or suppression of the secreting cells.

[1]In this example the sponge was (accidentally) exposed for a period of 24h in a staining experiment with Alizarine red S. Although no trace of the agent was discovered in the exposed sponge, it developed this type of deviating form.

Fig. 6.1. An example of a growth form of the sponge *Haliclona oculata*, in which the growth process was experimentally disturbed by a (toxic) agent

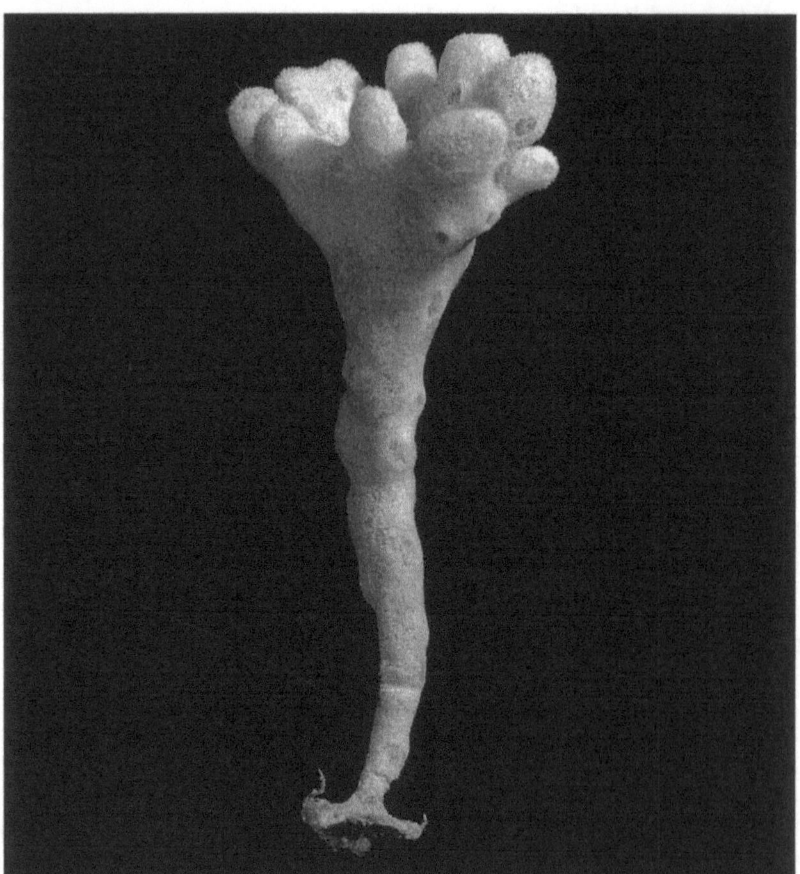

Fig. 6.2. Example of a cup-shaped growth form of the sponge *Haliclona oculata* as sometimes can be found in the field

Cup-shaped growth forms

The forms in Figs. 6.1 and 6.2 demonstrate that the radiate accretive growth process can result in a cup-shaped growth form: when the branches around the local minimum would anastomose a real cup is formed. This type of growth form is characteristic for many other sponge genera. These cup shapes can become very efficient hydrodynamically when a small feature of the aquiferous system is changed, where the inhalant pores (see Fig. 3.14) are situated on the outside of the cup and the exhalant ones at the inside.

Simulation of a community

Another application of morphological simulation models is to mimic the emergence of forms in a community of organisms. In all examples discussed so far, the growth forms as well as the simulated forms are assumed to be formed in isolation. In reality, for example in a coral reef, the forms do not develop as individuals. The individual forms will compete for space, light, nutrients, etc. Sessile organisms in a reef can chemically suppress each other and are arranged in a competitive network. These communities show a complex behaviour, which cannot be understood with simulation models of the individual composing elements. From these complex communities, as found in coral reefs, still little insight has been obtained. A morphological model, consisting of several of the dominant species and the physical environment, could be a way to provide more information on these reef communities.

References

1. W.F. Ames: *Numerical methods for partial differential equations.* Academic Press, New York 1977.
2. M. Aono and L. Kunii: Botanical tree image generation. *IEEE Computer Graphics and Applications* 8:10–34 (1984).
3. M.F. Barnsley: *Fractals everywhere.* Academic Press, Boston New York 1988.
4. D. Barthel: Influence of different current regimes on the growth form of *Halichondria panicea* Pallas. In: J. Reitner and H. Keupp, (eds.) *Fossil and recent sponges*, pp. 387–394, Springer, Berlin Heidelberg 1991.
5. A.D. Bell: The simulation of branching patterns in modular organisms. In: J. L. Harper, B. R. Rosen, and J. White, (eds.) *The growth and form of modular organisms*, pp. 143–159, The Royal Society London, London 1986.
6. P.J. Besl and R.C. Jain: Invariant surface characteristics for 3D object recognition in range images. *Computer Vision, Graphics and Image Processing* 33:33–80 (1986).
7. G.P. Bidder: The relation of the form of a sponge to its currents. *Quart. J. Microsc. Sci.* 67:293–323 (1923).
8. M.J.M. de Boe: *Analysis and computer generation of division patterns in cell layers using developmental algorithms.* PhD thesis, Rijks Universiteit Utrecht 1989.
9. J.P. Boon and A. Noullez: Development, growth, and form in living systems. In: H.E. Stanley and N. Ostrowsky, (eds.) *On growth and form*, pp. 174–183, Martinus Nijhoff, Boston 1987.
10. D.W.J. Bosence: Ecological studies on two unattached coralline algae from western Ireland. *Paleontology* 19(2):365–395 (1976).
11. R.H. Bradbury and R.E. Reichelt: Fractal dimension of a coral reef at ecological scales. *Mar. Ecol. Prog. Ser.* 10(2):169–172 (1983).
12. R.M. Brady and R.C. Ball: Fractal growth of copper electrodeposits. *Nature* 309:225–229 (1984).
13. P. Brien, C. Lévi, M. Sara, O. Tuzet, and J. Vacelet: *Traité de zoologie, anatomie, systématique, biologie, Tome III Spongiares, fascicule 1.* Masson et Cie Editeurs, Paris 1973.

14. L. Bunkley-Williams and E.H. Williams: Global assault on coral reefs. *Natural History* 4/90:46–54 (1990).
15. J.C. Bythell: A total nitrogen and carbon budget for the elkhorn coral *Acropora palmata* (Lamarck). *Proceedings of the 6th International Coral Reef Symposium* 2:535–540 (1988).
16. R.F. Carl and R.E. Smalley: Fullerenes. *Scientific American* 265(4):32–41 (1991).
17. J.M. Dauget: Application of tree architectural models to reef-coral growth forms. *Mar. Biol.* 111:157–165 (1991).
18. F.M. Dekking: Recurrent sets. *Advances in Mathematics* 44:78–104 (1982).
19. S. Demko, L. Hodges, and B. Naylor: Construction of fractal objects with iterated function systems. *Computer Graphics* 19(3):271–278 (1985). SIGGRAPH '85 Proceedings.
20. E.H. Dooijes and Z.R. Struzik: The practical measurement of fractal parameters. In E. Louis, L.M. Sander, P. Meakin, and J.M. Garcia-Ruiz, (eds.) *Growth patterns in physical sciences and biology*, pp. 111–120, Plenum Press, New York 1993.
21. G.D. Doolen: *Lattice gas methods for partial differential equations.* Addison-Wesley, New York 1990. Proceedings Volume IV, Santa Fe Institute, Studies in the Science of Complexity.
22. P.J. Edmunds and P. Spencer Davies. An energy budget for *Porites porites* (Scleractinia) growing in a stressed environment. *Coral Reefs* 8:37–43 (1989).
23. K.J. Falconer: *The geometry of fractal sets.* Cambridge University Press 1985.
24. K.J. Falconer: *Fractal geometry, mathematical foundations and applications.* John Wiley & Sons, New York 1990.
25. F. Family, B.R. Masters, and D.E. Platt: Fractal patterns formation in human retinal vessels. *Physica D* 38:98–103 (1989).
26. J. Feder: *Fractals.* Plenum Press, New York London 1988.
27. J. Feder, E.L. Hinrichsen, K. J. Maløy, and T. Jøssang: Geometrical crossover and self-similarity of DLA and viscous fingering clusters. *Physica D* 38:104–111 (1989).
28. L. De Floriani: A pyramidal data structure for triangle based surface description. *IEEE Computer Graphics and Applications* 9(2):67–78 (1989).
29. J.D. Foley, A. van Dam, S.K. Feiner, and J.F. Hughes: *Computer graphics: principles and practice.* Addison-Wesley, New York 1990.
30. D.R. Fowler, H. Meinhardt, and P. Prusinkiewicz: Modelling seashells. *Computer Graphics* 26(2):379–387 (1992).
31. L. Franzisket: Riffkorallen können autotroph leben. *Naturwissenschaften* 56:144 (1969).
32. D. Frijters: An automata-theoretical model of the vegetative and flowering development of *Hieracium murorum* L. *Biol. Cybern.* 24:1–13 (1976).

33. H. Fujikawa and M. Matsushita: Fractal growth of *Bacillus subtilis* on agar plates. *J. Phys. Soc. Japan* 58(11):3875–3878 (1989).
34. H. Fujikawa and M. Matsushita: Bacterial fractal growth in the concentration field of nutrient. *J. Phys. Soc. Japan* 60(1):88–94 (1991).
35. P.W. Glynn and L.D. Croz: Experimental evidence for high temperature stress as the cause of El Niño-coincident coral mortality. *Coral Reefs* 8:181–191 (1990).
36. A.L. Goldberger, D.R. Rigney, and B.J. West: Chaos and fractals in human physiology. *Scientific American* 262:34–41 (1990).
37. T.J. Goreau and A.H. Macfarlane: Reduced growth rate of *Montastrea annularis* following the 1987-1988 coral-bleaching event. *Coral Reefs* 8:211–215 (1990).
38. R.R. Graus and I.G. Macintyre: Variation in growth forms of the reef coral *Montastrea annularis* (Ellis and Solander): a quantitative evaluation of growth response to light distribution using computer simulation. *Smithson. Contr. Mar. Sci.* 12:441–464 (1982).
39. E. Haeckel: Report on the Radiolaria collected by H.M.S. Challenger. *Chall. Rep. Zool.* 18 (1887).
40. J. L. Harper, B. R. Rosen, and J. White: *The growth and form of modular organisms*. The Royal Society London, London 1986.
41. W.D. Hartman: Natural history of the marine sponges of southern New England. *Bull. Peabody Mus. Nat. Hist.* 12:1–155 (1958).
42. M. Hata: On the structure of self-similar sets. *Japan J. Appl. Math.* 2:381–414 (1985).
43. P. Hogeweg and B. Hesper: A model study on biomorphological description. *Pattern Recognition* 6:165–179 (1974).
44. J.E. Hopcroft and J.D. Ullman: *Introduction to automata theory, languages and computation*. Addison-Wesley, London 1979.
45. J.E. Hutchinson: Fractals and Self Similarity. *Indiana University Mathematics Journal* 30(5):713–747 (1981).
46. J.B.C. Jackson: Morphological strategies of sessile animals. In: C. Larwood and B.R. Rosen, (eds.) *Biology and systematics of colonial organisms volume II*, pp. 499–555, Academic Press, London New York 1979.
47. J.B.C. Jackson, L.W. Buss, and R.E. Cock: *Population biology and evolution of clonal organisms*. Yale University Press, New Haven London 1985.
48. D. Jebram: Influences of the food on the colony forms of *Electra pilosa* (Bryozoa, Cheilostomata). *Zool. Jb. Syst.* 108:1–14 (1980).
49. R.E. Johannes: Sources of nutritional energy for reef corals. *Proceedings of the 2th International Coral Reef Symposium* 1:133–137 (1974).
50. G. Johnston: *A history of British sponges and lithophytes*. Lizars, Edinburgh 1862.
51. P.L. Jokiel and S.L. Coles: Response of Hawaiian and other Indo-Pacific reef corals to elevated temperature. *Coral Reefs* 8:155–162 (1990).

52. G. Julia: Mémoire sur l'iteration des functions rationelles. *J. Math.* pp. 47–245 (1918).
53. J.A. Kaandorp: Interactive generation of fractal objects. In: G. Marechal, (ed.) *Proceedings of the European Computer Graphics Conference*, pp. 181–196, North-Holland, Amsterdam New York 1987.
54. J.A. Kaandorp: Modelling growth forms of sponges with fractal techniques. In: A.J. Crilly, R.A. Earnshaw, and H. Jones, (eds.) *Fractals and chaos*, pp. 71–88, Springer, New York Berlin 1991a.
55. J.A. Kaandorp: Modelling growth forms of the sponge *Haliclona oculata* (Porifera; Demospongiae) using fractal techniques. *Mar. Biol.* 110:203–215 (1991b).
56. J.A. Kaandorp: Simulating radiate accretive growth using iterative geometric constructions. In: E. Louis, L.M. Sander, P. Meakin, and J.M. Garcia-Ruiz, (eds.), *Growth patterns in physical sciences and biology*, pp. 331–340, Plenum Press, New York 1993a.
57. J.A. Kaandorp: 2D and 3D modelling of marine sessile organisms. In: A.J. Crilly, R.A. Earnshaw, and H. Jones, (eds.), *Applications of fractals and chaos*, Springer, New York Berlin 1993b.
58. J.A. Kaandorp: A formal description of radiate accretive growth. J. Theor. Biol. (in press).
59. J.A. Kaandorp and M.J. de Kluijver: Verification of fractal growth models of the sponge *Haliclona oculata* (Porifera; class Demospongiae) with transplantation experiments. *Mar. Biol.* 113:133–143 (1992).
60. A. Kaufman: Efficient algorithms for 3D scan-conversion of parametric curves, surfaces and volumes. *Computer Graphics* 21(4):269–277 (1987).
61. A. Kaufman: Efficient algorithms for scan-converting 3D polygons. *Comput. & Graphics* 12(2):213–219 (1988).
62. Y. Kawaguchi: A morphological study of the form of nature. *Computer Graphics* 16(3):223–232 (1982). SIGGRAPH '82 Proceedings.
63. M.J. de Kluijver: Sublittoral hard substrate communities of the southern Delta area, SW Netherlands. *Bijdr. Dierk.* 59(3):141–158 (1989).
64. M.J. de Kluijver and R.J. Leewis: Changes in the sublittoral hard substrate communities in the Oosterschelde estuary (SW Netherlands), caused by changes in the environmental parameters. To appear in: The Oosterschelde estuary: case study of a changing ecosystem. Ed. Nienhuis, P.H. & A.C. Smaal. Hydrobiologia.
65. M.A.R. Koehl: Mechanical design in sea anemones. In: G.O. Mackie, (ed.), *Coelenterate ecology and behaviour*, pp. 23–31, Plenum Press, New York 1976.
66. M.A.R. Koehl: The interaction of moving water and sessile organisms. *Scientific American* 247(6):110–120 (1982).
67. H. Koop: *Forest dynamics*. Springer, Berlin Heidelberg 1989.
68. G. Larwood and B.R. Rosen: *Biology and systematics of colonial organisms*. Academic Press, London New York 1979.

69. H.A. Lauwerier: *Fractals. Meetkundige figuren in eindeloze herhaling.* Aramith uitgevers, Amsterdam 1987.
70. H.A. Lauwerier and J.A. Kaandorp: Fractals (mathematics, programming and applications). In: M.M. de Ruiter, (ed.) *Advances in computer graphics III*, pp. 177–205, Springer, Berlin New York 1988.
71. J.B. Lewis: Banding, age and growth in the calcereous hydrozoan *Millepora complanata* Lamarck. *Coral Reefs* 9:209–214 (1991).
72. Y.S. Lim, T.I. Cho, and K.H. Park: Range image segmentation based on 2D quadratic function approximation. *Pattern Recognition Letters* 11:699–708 (1990).
73. A. Lindenmayer: Mathematical models for cellular interactions in development. *J. Theor. Biol.* 18:280–299 (1968).
74. E.A. Lord and C.B. Wilson: *The mathematical description of shape and form.* Ellis Horwood, New York 1984.
75. J. Lovelock: *The ages of Gaia.* Oxford University Press, Oxford 1988.
76. B.B. Mandelbrot: Fractal aspects of the iteration $z \rightarrow lz(1-z)$ for complex l and z. *Annals of the New York Academy of Sciences* 357:249–259 (1980).
77. B.B. Mandelbrot: *The fractal geometry of nature.* Freeman, San Francisco 1983.
78. B.B. Mandelbrot and C.J.G. Evertsz: The potential distribution around growing fractal clusters. *Nature* 348:143–145 (1990).
79. J.M. Marden: The continental shelf: man's new frontier. *National Geographic* 153(4):495–531 (1978).
80. M. Matsushita and H. Fujikawa: Diffusion-limited growth in bacterial colony formation. *Physica A* 168:498–506 (1990).
81. T. Matsuyama, M. Sogawa, and Y. Nakagawa: Fractal spreading growth of *Serratia marcescens* which produces surface active exolipids. *FEMS Microbiology Letters* 61:243–246 (1989).
82. P. Meakin: Diffusion-contolled cluster formation in 2-6 dimensional space. *Physical Review A* 27(3):1495–1507 (1983).
83. P. Meakin: Diffusion-controlled cluster formation in two, three and four dimensions. *Physical Review A* 27(1):604–607 (1983).
84. P. Meakin: A new model for biological pattern formation. *J. Theor. Biol.* 118:101–113 (1986).
85. H. Meinhardt and H. Klingler: A model for pattern formation on the shells of molluscs. *J. Theor. Biol.* 126:63–89 (1987).
86. H. Meinhardt and M. Klingler: Schnecken- und Muschelschalen: Modelfall der Musterbildung. *Spektrum der Wissenschaft* 8:60–69 (1991).
87. D.R. Morse, J.H. Lawton, M.M. Dodson, and M.H. Williamson: Fractal dimension of vegetation and the distribution of arthropod body length. *Nature* 314:731–733 (1985).
88. J.D. Murray: Generation of biological pattern and form. *IMA Journal of Mathematics Applied in Medicine & Biology* 1:51–75 (1984).

89. J.D. Murray: How the leopard gets its spots. *Scientific American* 258(3):62–69 (1988).
90. J.D. Murray: *Mathematical Biology*. Springer, Berlin Heidelberg 1990.
91. T. Nakamori: Skeletal growth model of the dendritic hermatypic corals limited by light shelter effect. *Proceedings of the 6th International Coral Reef Symposium* 3:113–118 (1988).
92. W.M. Newman and R.F. Sproull: *Principles of interactive computer graphics*. McGraw-Hill, London 1979.
93. L. Niemeyer, L. Pietronero, and H.J. Wiesmann: Fractal dimension of dielectric breakdown. *Phys. Rev. Lett.* 52(12):1033–1036 (1984).
94. J. Nittmann, G. Daccord, and G. H.E. Stanley: Fractal growth of viscous fingers: quantitative characterization of a fluid instability phenomenon. *Nature* 314:141–144 (1985).
95. M. Ohgiwari, M. Matsushita, and T. Matsuyama: Morphological changes in growth phenomena of bacterial colony patterns. To be published in *J. Phys. Soc. Japan*.
96. J.K. Oliver: Intra-colony variation in the growth of *Acropora formosa*: extension rates and skeletal structure of white (zooxanthellae-free) and brown-tipped branches. *Coral Reefs* 3:139–147 (1984).
97. H. O. Peitgen and P. H. Richter: *The Beauty of Fractals*. Springer, Berlin Heidelberg 1986.
98. J.W. Porter: Zooplankton feeding by the Caribbean reef-building coral *Montastrea cavernosa. Proceedings of the 2th International Coral Reef Symposium* 1:111–125 (1974).
99. W.H. Press, B.P. Flannery, S.A. Teukolsky, and W.T. Vetterling: *Numerical recipes in C.* Cambridge University Press, Cambridge 1988.
100. P. Prusinkiewicz and A. Lindenmayer: *The algorithmic beauty of plants*. Springer, New York Berlin 1990.
101. P. Prusinkiewicz, A. Lindenmayer, and J. Hanan: Developmental models of herbaceous plants for computer imagery purposes. *Computer Graphics* 22(4):141–150 (1988). SIGGRAPH '88 Proceedings.
102. P. de Reffye, C. Edelin, J. Francon, M. Jaeger, and C. Puech: Plant models faithful to botanical structure and development. *Computer Graphics* 22(4):151–158 (1988). SIGGRAPH '88 Proceedings.
103. E. Renshaw: Computer simulation of sitka spruce: spatial branching models for canopy growth and root structure. *IMA Journal of Mathematics Applied in Medicine & Biology* 2:183–200 (1985).
104. J.P. Rigaut: Fractals, semi-fractals and biometry. In: G. Cherbit, (ed.) *Fractals non-integral dimensions and applications*, pp. 151–187, John Wiley & Sons, New York 1991.
105. B. Rinkevich and Y. Loya: Oriented translocation of energy in grafted reef corals. *Coral Reefs* 1:243–247 (1983).

106. P.J. Roos: *Growth and occurrence of the reef coral* Porites astreoides *Lamarck in relation to submarine radiance distribution.* PhD thesis, University of Amsterdam 1967.
107. A. Rosenfeld and A.C. Kak: *Digital Picture Processing.* Academic Press 1976.
108. J.A. Rubin: Growth and refuge location in continuous, modular organisms: experimental and computer simulation studies. *Oecologia* 72:46–51 (1987).
109. L.M. Sander: Fractal growth processes. *Nature* 322:789–793 (1986).
110. L.M. Sander: Fractal growth. *Scientific American* 256(1):82–88 (1987).
111. F.J.A. Saris and T. Aldenberg: *Ecosysteemmodellen; mogelijkheden en beperkingen.* Pudoc, Wageningen 1986. Proceedings van de PSG studiedag over mathematische ecosysteem modellen, Den Haag 1986.
112. H.A. Simons: *The sciences of the artificial.* MIT Press, Cambridge MA 1969.
113. A.R. Smith: Plants, fractals and formal languages. *Computer Graphics* 18(3):1–10 (1984). SIGGRAPH '84 Proceedings.
114. R.W.M. van Soest: The Indonesian sponge fauna: a status report. *Netherlands Journal of Sea Research* 23(2):222–230 (1989).
115. R.W.M. van Soest and J. Verseveldt: Unique symbiotic octocoral-sponge association from Komodo. *Indo-Malayan Zoology* 4:27–32 (1987).
116. W.S. Sollas: Report on the Tetractinellida collected by H.M.S. Challenger during the years 1873–1876. *Rep. H.M.S. Challenger Scient. Results* 25:1–458 (1878).
117. P. Spencer Davies: The role of zooxanthellae in the nutritional energy requirements of *Pocillopora eydouxi. Coral Reefs* 2:181–186 (1984).
118. H.E. Stanley and N. Ostrowsky: *On growth and form: fractal and non-fractal patterns in physics.* Martinus Nijhoff, Boston 1987.
119. D.L. Taylor: Intra-colonial transport of organic compounds and calcium in some Atlantic reef corals. *Proceedings of the 3th International Coral Reef Symposium* 2:432–436 (1977).
120. D. Terzopoulos: Regularization of inverse visual problems involving discontinuities. *IEEE Transactions on pattern analysis and machine intelligence* PAMI-8:413–424 (1986).
121. D.W. Thompson: *On growth and form.* Cambridge University Press, Cambridge 1942.
122. D.L. Turcotte, R.F. Smalley Jr, and S.A. Solla: Collapse of loaded fractal trees. *Nature* 313:671–672 (1985).
123. A.M. Turing: The chemical basis of morphogenesis. *Phil. Trans. Roy. Soc. Lond.* B237:37–72 (1952).
124. B.W. Unger and D.S. Bidulock: Modular design of multicomputer systems. *Simulation* 7:1–9 (1981).

125. J.E.N. Veron and Pichon: *Australian Institute of Marine Science Monograph series vol 1, Scleractinia of eastern Australia part 1.* Australian Government Publishing Service, Canberra 1976.
126. A.D. Vethaak, R.J.A. Cronie, and R.W.M van Soest: Ecology and distribution of two sympatric, closely related sponge species, *Halichondria panicea* (Pallas, 1766) and *H. bowerbanki* Burton, 1930 (Porifera, Demospongiae), with remarks on their speciation. *Bijdragen tot de Dierkunde* 52(2):82–102 (1982).
127. S. Vogel: Current-induced flow through the sponge, *Halichondria. Biol. Bull.* 147:443–456 (1974).
128. S. Vogel: *Life in moving fluids.* Princeton University Press, Princeton 1983.
129. S. Vogel and W.L. Bretz: Interfacial organisms: passive ventilation in the velocity gradients near surfaces. *Science* 175:210–211 (1971).
130. T. Vuorisalo and J. Tuomi: Unitary and modular organisms: criteria for ecological division. *Oikos* 47(3):382–385 (1986).
131. S.A. Wainwright and M.A.R. Koehl: The nature of flow and the reaction of benthic cnidaria to it. In: G.O. Mackie, (ed.) *Coelenterate ecology and behaviour*, pp. 5–21, Plenum Press, New York 1976.
132. F.E. Warburton: Influence of currents on form of sponges. *Science* 132:89 (1960).
133. W.H de Weerdt: Transplantation experiments with Caribbean *Millepora* species (Hydrozoa, Coelenterata), including some ecological observations on growth forms. *Bijdragen tot de Dierkunde* 51(1):1–19 (1981).
134. W.H. de Weerdt: A systematic revision of the north-eastern Atlantic shallow-water Haplosclerida (Porifera, Demospongiae), part II: Chalinidae. *Beaufortia* 36(6):81–165 (1986).
135. M.J. Wenninger: *Polyhedron models.* Cambridge University Press, Cambridge 1971.
136. M.J. Wenninger: *Spherical models.* Cambridge University Press, Cambridge 1979.
137. F. Wiedenmayer: *Shallow-water sponges of the western Bahamas.* Birkhäuser, Basel 1977.
138. M. Wijkstra: Volume modeling. Technical report, University of Amsterdam, Faculty of Mathematics and Computer Science 1991. MSc thesis.
139. C.R. Wilkinson, A.C. Cheshire, D.W. Klumpp, and A.D. McKinnon: Nutritional spectrum of animals with photosynthetic symbionts-corals and sponges. *Proceedings of the 6th International Coral Reef Symposium* 3:27–30 (1988).
140. T.A. Witten and L.M. Sander: Diffusion-limited aggregation, a kinetic critical phenomenon. *Phys. Rev. Lett.* 47(19):1400–1403 (1981).
141. P. Wlczek, H.R. Bittner, and M. Sernetz: 3-dimensional image-analysis and synthesis of natural fractals. *Acta Stereol* 8(2):315–324 (1989).
142. S. Wolfram: Cellular automata. *Rev. Mod. Phys.* 55(601) (1983).

Index

Springer-Verlag and the Environment

We at Springer-Verlag firmly believe that an international science publisher has a special obligation to the environment, and our corporate policies consistently reflect this conviction.

We also expect our business partners – paper mills, printers, packaging manufacturers, etc. – to commit themselves to using environmentally friendly materials and production processes.

The paper in this book is made from low- or no-chlorine pulp and is acid free, in conformance with international standards for paper permanency.

MIX
Papier aus verantwortungsvollen Quellen
Paper from responsible sources
FSC® C105338

If you have any concerns about our products,
you can contact us on
ProductSafety@springernature.com

In case Publisher is established outside the EU,
the EU authorized representative is:
Springer Nature Customer Service Center GmbH
Europaplatz 3, 69115 Heidelberg, Germany

Printed by Libri Plureos GmbH
in Hamburg, Germany